New Technologies for Smart Grid Operation

New Technologies for Smart Grid Operation

Sioe T Mak

PhD, EE, IEEE Life Fellow, Senior Consultant
ESTA International, LLC, USA

IOP Publishing, Bristol, UK

ISBN 978-0-7503-1158-8 (ebook)
ISBN 978-0-7503-1159-5 (print)
ISBN 978-0-7503-1160-1 (mobi)

DOI 10.1088/978-0-7503-1158-8

Version: 20150201

IOP Expanding Physics
ISSN 2053-2563 (online)
ISSN 2054-7315 (print)

British Library Cataloguing-in-Publication Data: A catalogue record for this book is available from the British Library.

Published by IOP Publishing, wholly owned by The Institute of Physics, London

IOP Publishing, Temple Circus, Temple Way, Bristol, BS1 6HG, UK

US Office: IOP Publishing, Inc., 190 North Independence Mall West, Suite 601, Philadelphia, PA 19106, USA

Contents

Preface

This book is intended to provide a comprehensive overview of the requirements for implementing advanced smart grid applications. Advances in communication technology, satellite communication, digital electronics, database management systems and computer technology allow the development of applications that were once deemed impossible to implement. There is a synergism between meter reading and smart grid applications and electric utility communication infrastructure design. Their effects on the operation of communication and energy delivery infrastructure are discussed extensively in this book. In addition, the ability to introduce synchronization time clocks through the now available phasor measuring unit (PMU) technology adds a new dimension to the functional capability of smart grids.

Much of the content of this book is based directly on the author's previous experience in developing and implementing power line two-way communication technology for load management and demand response applications. The author has also developed methodologies for using the same communication technology in implementing advanced meter reading applications and some distribution automation functions in the USA and other countries. Discussions with electric utility engineers, vendors and manufacturers of sensors and distribution equipment, and authors and scientists from various institutions involved with smart grid studies and field investigations have also provided a wealth of information and material for this book.

The comparison of different communication technologies for electric utility applications is also extensively discussed in this book. A chapter is also devoted to issues looming over the horizon due to the increased amount of distributed generation at the distribution level. The impact of charging systems for electric vehicles on the metrology and rate design, demand control and network operation of the electric energy delivery infrastructure are also addressed.

A bibliography of published papers, vendors and other reference material is included at the end of the book.

Acknowledgement

I like to dedicate this book to the engineers and scientists who dedicate their research and time to improving the operation of our electric power industry. Also, I am grateful for the encouragement and support from my wife and daughters while I was writing this book. In particular I would also like to thank the reviewers and the editorial and production staff of IOP Publishing for their excellent efforts and accurate reviews of this book.

Author biography

Sioe T Mak

Sioe T Mak received his Diploma in Electrical Engineering from the University of Indonesia, the MSc and PhD in EE from the Illinois Institute of Technology in Chicago. He was Senior Staff Scientist at the Distribution Control Systems, Inc., a subsidiary of ESCO Technologies Corp. He served on many IEEE Committees, published numerous papers in the area of EHV Pole Fires, High Voltage Insulation Contamination and Power Frequency Communication Technology. He is also involved with Communication Infrastructure Studies, Reliability and Life of Electric Equipment, Remote Metering and Distribution Automation and Power Quality. He holds US and world-wide patents in the area of PLC technology (TWACS), currently used in the USA by many electric utilities for AMI and Distribution Automation. He is also an IEEE Life Fellow.

IOP Publishing

New Technologies for Smart Grid Operation

Sioe T Mak

Chapter 1

The future of electric power systems

1.1 General introduction

Currently, many electric utility systems generate bulk electric power by converting energy from different fossil fuels or nuclear material into heat, which is subsequently used to generate steam power for operating machines which drive electric generators to produce electric energy. Hydropower is another source used to generate electric power. A network of transmission lines is used to transport the electric energy from the generating plants to population centers and a massive distribution network distributes the energy to users. At the user end, voltages are dropped to a safe level and many different devices convert the electric energy into a form of energy for the immediate use of the customers. During the latter part of the 20th century, wind and solar power conversion devices made their mark as another possible source of electric power.

Electric energy is the life-blood of our global civilization and is here to stay. The demand for electricity continues to increase and much effort is devoted to optimizing its delivery, conversion and use. With an aging energy delivery infrastructure and an increase in demand, much effort is aimed at improving the life, operational reliability and efficiency of existing systems to meet the present and future growth in the number of customers and the associated service requirements. The concept of a smart grid has recently been conceived and promoted by the electric power industry, engineers and various government agencies to enhance the operation and reliability of power generation and delivery to customers. This development has been partially triggered by environmental compliance considerations and also by new concepts of demand response. The demand response concept was at one time equivalent to the load management function which was triggered by the need to reduce peak demand to avoid using standby peaking unit generating plants. Demand response now has a completely different meaning. Customer participation in reducing peak demand, supporting asset management functions to reduce system losses and carbon emissions into the atmosphere, etc, are just a few of the possibilities considered by the

doi:10.1088/978-0-7503-1158-8ch1

industry. In addition, the proliferation of plug-in hybrid electric vehicles (EVs) and the distributed generation of renewable resources coupled with storage capabilities, have started to pose new challenges for system upgrades and operational reliability. The transition to smart grids will be at best evolutionary in order not to disrupt the continuity of service to the customers.

Bulk electric power can be generated at sites away from populated areas, making it more amenable for pollution control, reducing safety and health hazards, etc. Distributed generation (wind power, low head hydro, etc) is being increasingly installed as the price drops and tax incentives are given by governments to encourage the generation of renewable green power.

In many states in the USA, local state governments have started to impose renewable portfolio standards which are primarily driven by measures to limit greenhouse gases, demand response, energy conservation and environmental issues. Electric supply companies are required to provide a minimum percentage of their electricity from approved renewable energy sources. Also, according to the policy statement of the Federal Regulatory Commission, demand response can:

- provide competitive pressure to reduce wholesale power prices;
- increase customer awareness of energy usage;
- provide for more efficient operation of markets;
- mitigate market power;
- enhance reliability;
- in combination with certain new technologies support the integration of renewable energy resources, distributed generation and advanced metering technologies.

In order to obtain a better understanding of how these capabilities can be implemented in economical and gradual comprehensive steps without causing undue interruptions to the energy supply and delivery to customers, the following sections of this chapter provide physical insight into what electric energy is and how current electric power systems operate to deliver this energy to customers.

1.2 The electric energy delivery infrastructure

Electric energy is unique because it cannot be used in its physical form to meet our needs. It is a transitional form of energy which can be easily manipulated and transported. When one mentions ‘electric energy’, one is basically dealing with the generation, transportation, distribution and delivery of energy. Electric energy is the most versatile transitional form of energy with the following desirable basic characteristic features.

It can be transported along narrow corridors in large bulk quantities at practically the speed of light. These narrow transport corridors are the transmission lines, the distribution network, etc. Also, to serve a multitude of energy users scattered over wide geographical areas, multiple independent, parallel and simultaneous energy transport channels are used. These channels form an energy distribution network, which is basically the delivery infrastructure. For an electric distribution network, the major components of the delivery infrastructure are the distribution substation buses from which feeders emanate in different directions and where step-down transformers are used to provide power to the service voltage network. The most common form used today is alternating current (ac) operating at 50 Hz and 60 Hz. This allows the use of transformers to convert the voltage into various levels as required for a safe user level. Direct current (dc) is still used, but only for high voltage bulk power transmission over long distances.

Other unique properties of electric energy are:

- It can easily be split into minute quantities or bulk quantities, depending on the need.
- Conversion into other forms of energy that are useful to people. This is a well-known technology. Conversion into thermal energy is accomplished by using electric resistance wires. Electromechanical conversion devices, such as electric motors, convert the electric power into useful mechanical power. Numerous other applications could be listed.

Most utility networks can still be viewed as having the following major components.

- **Major power conversion plants.** Plant capacities range between a few hundred MW and a few thousand MW. There are now more than 15 000 generators in 10 000 power plants in the USA.
- **A transmission network.** Transmission voltages range between 34.5 kV and 765 kV in the USA. Currently, there are 170 000 miles of high voltage transmission lines rated at voltages higher than 200 kV.
- **A distribution network.** Distribution voltages range between 4.0 kV and 34.5 kV and at the service level voltages range between 120 V and 480 V. With nearly 6 million miles of lower voltage distribution lines, the US grid serves 125 million residential customers, 17.6 million commercial customers and 775 000 industrial customers. At the time of writing, the electric usage by the three groups of customers is, respectively, 37%, 36% and 27% of total generation.

1.3 Delivery network configurations and unique properties of the three-phase network

Three-phase networks operating at 50 Hz or 60 Hz still dominate the existing delivery infrastructure and the circuit configurations can be three-wire, three-wire with grounded neutral return, two-wire or single wire with earth return. The transitions to the various voltage levels are accomplished through the use of three-phase Y–Y with grounded or ungrounded neutrals, D–Y or single-phase transformers. Under steady state operating conditions, the phase to neutral voltages V_{an}, V_{bn} and V_{cn} are quite often transformed into line-to-line voltages on the secondary side by means of three-phase transformer winding configurations. However, when one views each of the phase to neutral voltages at the medium voltage substation bus as base phasors, and the line-to-line voltages as linear combinations of the base phasors, then a unique picture emerges showing some basic characteristic features of the

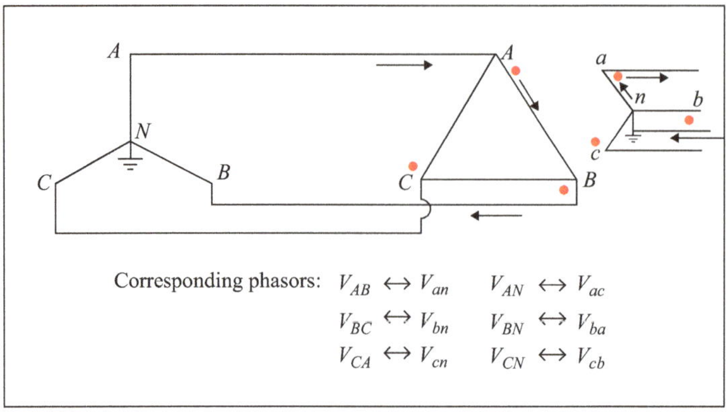

Figure 1.1. Diagram depicting the corresponding phasors rule.

electric network. For any reference phasor generated at the substation bus, there is a remote corresponding phasor at part of the network served by the substation that is slightly phase shifted with respect to the medium voltage substation bus reference phasor. Its magnitude depends on the circuit voltage drop and the intervening transformer winding ratio. A simple example of this is shown in figure 1.1.

Since the voltage is the source of the load current, the load current phasor can also be referenced with respect to the voltage source phasor. If there are no intervening power conditioning devices, one can make this phasor rule connection all the way from the remote customer end to the generator. Any current drawn on the 120 V side has its corresponding phasor on the generation side. Hence, if for any reason a massive cumulative imbalance of load occurs on the distribution side, it will also reflect itself as a massive imbalance on the generation side.

To improve the delivery of energy to customers, a variety of medium voltage and service voltage distribution network configurations are used. For rural areas, radial three-phase feeders served by a medium voltage substation bus prevail in the USA. In metropolitan areas, in particular in downtown areas, an underground low voltage network forms a massive interconnected grid served by a large number of step-down transformers. These transformers are sourced from several medium voltage feeders which obtain their power from a single transmission line or several transmission lines. Open loop medium voltage distribution lines are also common in many metropolitan areas. Most of the distribution transformers in the USA range in capacity between 25 kVA and 500 kVA. In other countries, distribution step-down transformer sizes in the MVA range serving a few hundred customers are not uncommon.

Distribution networks in particular have a unique property. Since there are no distribution circuits that have lengths close to a quarter wavelength of a 50 Hz or 60 Hz wave, no long line or Ferranti effects are expected to occur in the distribution network. This unique behavior of the phasors mentioned above has important implications for the design of control algorithms and applications which will be discussed in the following sections.

1.4 The electric power transient response behavior and energy conversion devices

Another important aspect is the time element, which is intimately linked to the transmitted and observed data. This involves the transient behavior of the network and how it affects the speed of response of a control action. When a switching action takes place, whether due to man-made control actions or naturally induced actions such as a lightning flashover across two conductors or a line fault, the electric network reacts according to some specific patterns.

Energy storage devices such as capacitors (stray capacitances or capacitor banks), inductors, etc, in the distribution network cause a reaction to a sudden change similar to the reaction of a physical system to a perturbation function. In most cases, the reaction in an electric power delivery network is transient oscillatory and decays within a very short time period, in general in less time than the duration of a quarter wavelength of the power frequency. The magnitude also depends on the part of the waveform of the power frequency voltage at which the switching action is initiated and the attenuation, which can be attributed to the energy absorbing parts of the network.

Figure 1.2 shows the transient oscillatory behavior of switching responses extracted from the voltage at the service voltage due to a switching operation at a medium voltage substation bus. A switching device is turned on at a certain angle before the voltage zero crossing of the medium bus phase conductor, causing a current flow into a current limiting load. Subsequently the switch is opened and remains open at a few degrees past the voltage zero crossing. The short duration current loop drawn by the load generates a transient oscillatory current at the feeder phase conductor causing a transient oscillatory voltage drop in the medium voltage feeder phase. This transient oscillatory voltage drop is extracted at the service voltage part of the circuit and is shown in the figure by the darkened waveform. The duration of the oscillatory response is longer than the load current pulse at the medium voltage bus but decays in less than half a cycle of the power voltage.

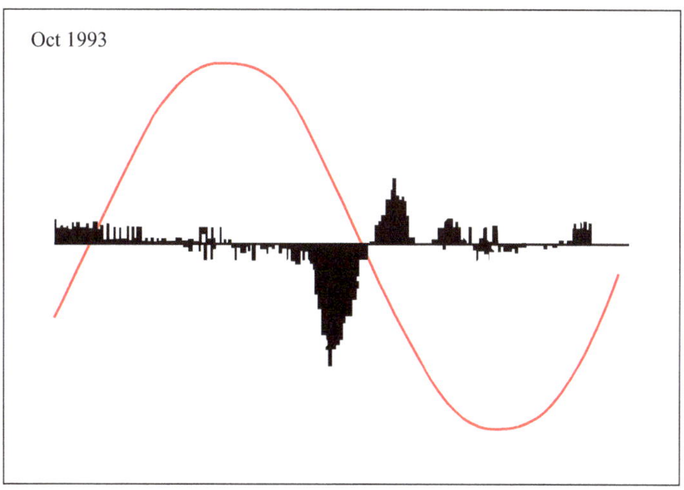

Figure 1.2. Extracted perturbation from field data. Perturbation is introduced at the 25 kV substation bus.

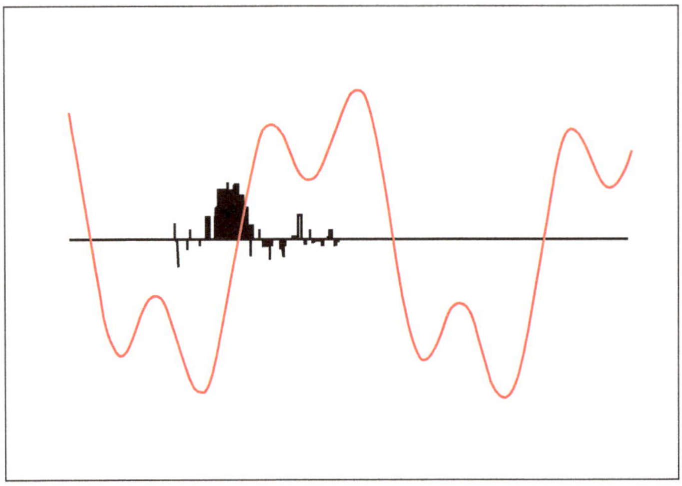

Figure 1.3. Signal captured at the neutral of a WYE distribution substation transformer.

Figure 1.3 shows an extracted current pulse from the neutral current transformer at a substation due to a large single pulsating load at the 240 V service voltage level.

Energy conversion devices such as electric motors, heaters, electric welders, etc, have relatively long transient time constants, with respect to the time it takes for one cycle of the power frequency to reach steady state operating conditions. The time constants typically last for multiple cycles of the 50 Hz or 60 Hz power frequency. An example is the locked rotor starting current of a motor as its output torque gradually increases. Electric ovens and space heaters show similar characteristics before they reach steady state temperatures.

1.5 Advanced metering and smart grid applications

This book primarily deals with electric utility distribution networks and how to use the available tools to optimize network operation and the use of energy by customers. Many of the new applications lend additional support and capabilities to already installed functions.

Lately, there has been a lot of discussion of new applications that will improve the utility network operation and the services to the utility customers. Advanced metering and smart grid functions, made possible by modern digital electronic devices, communication technologies and database technologies, are now being considered in many electric utilities and can be grouped in several main categories. A possible method of categorizing the applications is shown below. This list is by no means exhaustive. It is important to note that many of these functions and applications cannot be viewed or discussed independently without taking into account the synergism amongst them.

1.5.1 Advanced metering and demand response applications

- Electric energy consumption metering.
- New rate structures.

- Remote service connect and disconnect.
- Gas and water consumption metering.
- Load management/demand response.
- Customer services.
- Alarm and customer alarm applications.
- Prevention of unauthorized use of electricity.

1.5.2 Improvement of service reliability and optimization of energy delivery (asset management)

- Outage management and system restoration.
- Feeder load balancing and loss management.
- Integrated voltage and VAR control.
- Power quality monitoring.
- High impedance fault detection.
- Distributed generation integration.
- Hybrid EV and EV charging systems and their demand control.

1.5.3 Supporting functions

- Communication network monitoring and control.
- Extension of supervisory control and data acquisition (SCADA) capability into the distribution network.
- Integration of phasor measuring unit (PMU) technology into smart grid applications.
- AM/FM systems.
- Data management.
- Multi-party users.
- Etc.

1.6 Introduction of the time element

The ability to obtain synchronized, time-stamped interval metering data opens the door for numerous new applications which will benefit electric utilities and their customers. In metering one should also include all energy related data such as voltage, power quality, etc, which the new digital electronics smart meters can deliver. The duration of the monitoring interval depends on the application and the use of the collected information. For a total energy metering device such as a kWh meter, if the reading at time T_1 is P_1 kWh and at the time T_2 the reading is P_2 kWh, then $(P_2 - P_1)$ is the total energy consumed during the interval Δ, where $\Delta = (T_2 - T_1)$. The smaller the time interval Δ is chosen to be, the more accurately one can characterize the load behavior as a function of time. Since P_1 and P_2 both represent integrated values and not instantaneous magnitudes, all fluctuations of the energy consumption as a function of time are absorbed by the integration operation. Hence $\{(P_2 - P_1)/\Delta\}$ represents the average power consumed during the time interval Δ. For load survey and demand metering, the value of Δ is typically 15 min and for advanced billing and novel rate design applications Δ is typically 30 or 60 min.

If a voltage V_1 is read at time T_1, then this is normally assumed to be the rms voltage at T_1. From a mathematical standpoint, the rms value is the result of an integration process to obtain an equivalent dc average voltage for the duration of at least one cycle of the power frequency. The type of information one needs about the behavior of the voltage will determine what is collected, processed and stored within the time interval $\Delta = (T_2 - T_1)$. For voltage profiling of a feeder, the average value of the voltage for the time interval Δ and time stamped when the measurement was performed, is more important than the short time excursions of the peak values of the voltage within the time interval Δ. Time-stamping all of the data obtained, where the start and stop times of sampling (T_1 and T_2) are synchronized, is essential for time correlation of the gathered data with other events that may occur in the distribution network. The information about the energy state of the distribution network in relation to the locations of the energy users in the distribution grid will provide insight into where problems are or might arise. Remedial actions can be taken to prevent damage or nuisance problems in a timely fashion.

In the USA, the National Bureau of Standards cesium atomic clock in Boulder, Colorado, is used for time synchronization. The utility communication network has to broadcast this synchronizing time to all the remote units to set their time. A small variation of a few seconds in the times received by each remote unit, due to communication processing times and propagation delay, can be tolerated.

The hourly time-stamped readings provide sufficient accuracy for apportioning energy usage for billing purposes. A database warehousing all the time-stamped interval metered data is necessary so they can be made available to different interested parties.

In the transmission grid, the situation is completely different. The number of load centers (the transmission substations), ranging in size from a few tens of MW to a few hundred MWs, is small and they are most likely interconnected with each other. The distance between each of these load centers is hundreds of miles. More accurate data on the voltages and currents, correlated to each other in time and by location on the transmission grid, are needed. System operators should quickly be aware of any anomalous system behavior caused by disturbances in order to react accurately and in an expeditious fashion. Coherent real-time information about the voltage, current and other data according to a common time reference is needed. The introduction of satellite communications technology for GPS systems, which also broadcasts time information to Earth, provides a means for time synchronization of voltage, current, etc, monitoring at the various transmission load centers scattered over wide geographical areas. PMUs are now available for obtaining precise grid measurements at high speeds (typically 30 times per second and time-stamped) in a synchronous fashion.

The era of the smart transmission grid has arrived and it is time to integrate this new technology with the smart grid, which was initially conceived for implementation at the distribution level only. The link between the two major grids is the SCADA system, which will play a major role in the integration process. This topic will be discussed in the following chapters of this book.

IOP Publishing

Sioe T Mak

Chapter 2

Advanced metering infrastructure (AMI) and general operational and communication issues

2.1 Introduction

Recently many utilities have been aggressively implementing large scale automatic meter reading (AMR) capabilities in their systems. With the recent advances in computers, communication technologies and miniaturization of digital electronics devices, automation of remote meter reading has been made possible. Many of the utilities' concerns have initially involved the ability to read customer consumption of electric energy, gas and water on a scheduled basis for billing purposes only and possibly to perform some type of demand side management function.

However, several other reasons are also given to justify investing in a communication system for automated remote meter reading. Replacing meter readers is definitely not one of the main reasons. The annual cost of meter reading is very low in many areas. With a communication system in place, it is expected that other functional capabilities can be added that will benefit the utility company and its customers. The concepts of advanced metering infrastructure (AMI), advanced metering, demand response, smart meters, smart grids, distribution automation, utility assets management, improving power quality, home automation networks, etc, have started to appear in the technical literature, implying the evolution of futuristic systems. One common mistake is making a decision as to which communication technology to choose without a full understanding of which functions or tasks one is trying to accomplish using the technology. The choice of technology has to be driven by the functions and applications one expects to implement now and in the future for which the communication technology serves as one of the enabling tools to support the intended applications.

The trend in many countries, including the USA, of deregulating the electric utility industry through unbundling of the utility's monopolies, results in the creation of *generation*, *transmission* and *distribution* companies. This sets the stage for

doi:10.1088/978-0-7503-1158-8ch2

introducing numerous innovative concepts that are purported to allow competition and which will greatly benefit the energy users. The concept of *retail wheeling* at the distribution level generates new requirements for meter reading, customer billing, pricing strategies, etc. Electric energy is starting to lose its social function characteristics and is slowly being considered as a commodity. Non-electric retailers are exploring the possibilities of becoming energy retailers and several non-utility billing companies will eventually come into existence. The technologies are now available to make this possible.

Large scale AMI for remote meter reading capability implies the following possibilities:

- a reliable two-way communication system;
- the communication system can reach practically every customer connected to the utility's network;
- metering and other types of data can be brought back reliably and accurately in a timely fashion;
- a large number of data transfers per unit time can be handled by the communication network and processed for storage;
- storage of very large amounts of data organized for easy access by multiple interested parties;
- the installation or replacement of remote communication devices, which interface with the existing or new electric metering devices, should be easy and low in cost.

For the reasons mentioned above, electric utilities have started actively looking at what added-value capabilities can be implemented. The question that has been raised repeatedly is 'What else is possible beyond simple meter reading applications?'. Some applications which have been proposed are advanced metering, new types of billing systems, providing meter reading services to other utilities such as gas and water utilities, load control, time-of-use rates application, demand response management, remote service connect and disconnect, capacitor bank control, real-time voltage monitoring, outage management and other distribution automation functions. Expressions, such as smart meters, smart grids, asset management, data warehousing and metering data mining are becoming popular buzz words.

It is also obvious that the different parties within a utility organization have different applications in mind for utilization of the communication system. The customer services department desires timely retrieval of energy metering data for billing purposes. In addition to the regularly scheduled operation of the system for billing purposes, there are numerous possible unscheduled uses. Satisfying individual customer billing complaints, disconnecting services, pre-pay metering, customer profiles of energy use, etc, are just a few of these. The load management group primarily uses the one-way global outbound capability of the system to curtail total peak demand. Some preliminary load research requires load data, which can be obtained from time-stamped synchronized interval metering data. Outage management is a typical unscheduled application. These are only few of the many possible uses of the communication systems being considered. With the recent introduction of smart meters, more useful information can be collected and generated which can be used to design control algorithms that will benefit the operation of electric utilities. When added-value applications are as important as the meter reading function to the prospective user of the communication system, it is vital that all the important questions are asked and answered properly to each party's satisfaction. The dilemma faced has always been which questions to ask and which answers are good answers.

The topic that has to be addressed is the following. Can a system that is designed for utility AMR be economically expanded or upgraded to implement future added-value capabilities without requiring a major overhaul or a large increase in capital expenditure and future added utilization costs?

A communication system that is used to perform simple meter reading for monthly billing purposes may have severe limitations in terms of expandability to perform other added-value applications. The assessment of whether a technology will be able to meet future needs not only requires a good understanding of the strengths and limitations of the communication technology being considered, but also requires an intimate knowledge of the operational issues and expected system performance to handle the added-value functions. All these issues will be discussed in great detail in subsequent chapters.

Many communication systems used for utility applications are hybrids. The technology used for communication between a central computer and certain major gateway nodal points can be completely different from that used for communication between the major nodes and remote devices.

A typical system has a *central AMI communication server* (figure 2.1) which handles all communications to and from the remote intelligent devices and

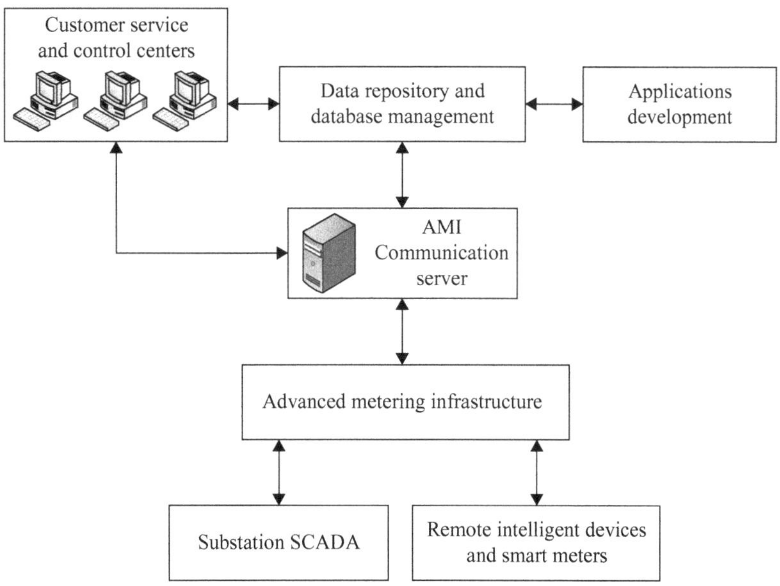

Figure 2.1. General system architecture for SMART GRID.

maintains the necessary database of communication paths and nodes, node and remote device addresses, and routing algorithms for efficient operation to access the remote devices. This computer is linked through a high-speed data link network to *primary intelligent nodes*. These nodes are the gateways to links with remote devices such as smart meters, demand controllers, distribution monitoring and control devices, and supervisory control and data acquisition (SCADA) substation devices. The server is also linked to other parties' service computers, such as the *customer service billing* computer, the *load management dispatch center* computer, the *service and maintenance dispatch center* computer, etc. The high-speed data link backhaul network can be a public telephone system, a cellular radio network, a satellite communication network, microwave radio, a fiber optic network, etc. The link from the primary gateway nodes to the remote devices can be a different communication network. This is normally owned by the utility and is solely dedicated to utility applications. For power line carrier types of communication technologies, the communication infrastructure is identical to the utilities' distribution power delivery infrastructure. Extensions of this network using hard-wired or radio technologies for reading water and gas meters are already available. Subnodes may or may not be required between the primary nodes and the remote metering, monitoring and control devices. These additional nodes can be repeaters or data concentrators. To simplify the description of the remote devices we use the generic name 'transponder' for all intelligent remote communication devices which are integrated with energy metering and other data collection devices, intelligent monitoring and control devices, alarm and switching devices, etc.

In this chapter, the substation SCADA is introduced and incorporated as part of the group of remote intelligent devices using a special link to the system's utility communication infrastructure. This addition opens the door to a vast number of new applications that benefit the operation of utilities. The following chapters include more detailed discussions of the role of SCADA and its importance in implementing new capabilities.

2.2 General communication and operational issues

If the metering data are only needed for scheduled billing of customers, then the requirements imposed on the communication system are not very stringent. Multiple repeated efforts to try to obtain a meter reading because of poor communication performance can be tolerated. If time interval metering data are required, more stringent requirements are imposed and even more so if the interval reading of all the meters has to be synchronized in time and time-stamped. The energy dispatch center, the rate designers, the demand side management group, the distribution control department and the billing department may be interested in the 15 min, half hourly or hourly synchronized data retrieved from all the electric energy metering devices.

This seemingly simple addition of the time element and time synchronization as a requirement triggers several levels of operational complexity. The economy of time utilization for data transfer, data throughput, communication reliability, data storage, quickly identifying changes in data transfer paths or link failures and their restoration are critical aspects to consider for the system.

In addition to the metering devices, there are load control devices, distribution control and monitoring devices, etc, and these are linked to a control or dispatch center through a backhaul communication network. From an operational standpoint there can be pre-programmed scheduled operations requiring information retrieval or simple switching type functions. Simple meter reading functions and load control belong to this category of operation. Unscheduled operations such as handling customer services, distribution network control operations, etc, also require the use of the communication system.

To assess whether a technology will be able to meet future needs requires a good understanding of the strengths and limitations of the communication technology being considered. An intimate knowledge of the operational issues and expected system performance to handle the added-value functions is also required.

The physical network and communication technology used for the infrastructure determine the flexibility and complexity of the operation and expansion of implementing future applications. A communication system is intended to transfer information from one point to another point in the most reliable, optimal and timely fashion. A tendency to define this statement in terms of ‘baud rate’, ‘being able to receive every bit reliably’, a ‘guarantee of obtaining all information 100% of the time’, etc, can be quite misleading. Figure 2.2 is used as an example. Suppose a system can communicate at a reasonably high baud rate of 10 000 bits s^{-1}. Assume that each message is 100 bits long. At 100% communication performance, 100 messages can be transmitted per second when operating in sequential fashion. For 1 000 000 messages to be retrieved, the time required will be 1 000 000/100 = 10 000 s which is approximately 2 h and 48 min.

If two different physical communication paths can be used simultaneously, and the two physical paths are *non-interfering*, then all the messages can be transmitted

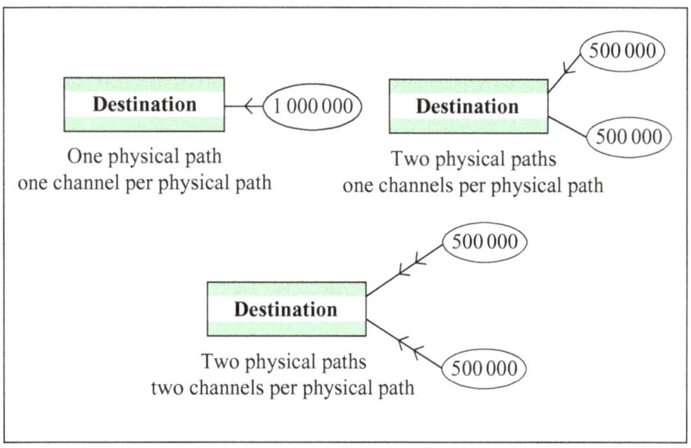

Figure 2.2. Transmission time reduction utilizing 2 channels on 2 non-physically interfering paths.

in 1 h and 24 min. If two *independent communication channels* can use the same path, as shown in the lower half of figure 2.2, then the transmission time can be reduced to 42 min.

Communication path interference, sometimes called cross-talk, behaves differently for the different types of communication technology used. As an example, power line carrier systems usually have their outbound transmitter and inbound receiver installed at the medium voltage distribution substation. Outbound signals intended for remote devices served by the distribution substation network quite often spill over to other medium voltage substations served by the same transmission line. These spill-over signals might interfere with ongoing communication activities at the other adjacent substations and mutually destroy each other. The only way to overcome this problem is by operating the communication activity in a sequential fashion for the substations served by the same transmission line. Hence concurrent communication activities at these substations are not possible. Simultaneous operation might be possible if the distribution substations are served by different transmission lines.

For radio frequency (RF) based communication infrastructure, the main transmitters and receivers can be located anywhere within the geographical area served by the electric utility distribution network. Repeater stations are scattered throughout the network and similar issues of mutual interference of communication channels apply here as well.

Simultaneous multichannel designs using the same physical communication path are already available and have been implemented by different manufacturers. Most of the designs use binary encoded messaging methods to transmit intelligence and use orthogonal design techniques to create simultaneous multichannel capabilities.

The example in figure 2.2 shows how system architecture and infrastructure design are intertwined. In the first design the 'hourly interval meter reading' requirement cannot be met. By the time the first meter is read again, 2 h and 48 min have elapsed due to communication delay time. The two parallel physical paths design allows double the flow of messages for the same amount of time. The design of the system almost meets the hourly meter reading requirement. These simple examples easily show that baud rate, when not coupled with system architecture and infrastructure design, does not provide all the information that the prospective user needs.

There is another level of complexity. Assume that only 90% of the data are perfectly received and 10% of the messages are lost because of communication performance problems:

- during the first communication try, 2 h and 48 min are used;
- during the second try about 17 min are needed to retrieve data from the remaining 10%;
- during the third try an additional 1.7 min are needed.

High communication performance at the first try, coupled with multiple non-interfering physical communication paths and the ability to use simultaneous multi-channel communication are the most desirable features to look for.

Another issue to consider is how many communication retries can be tolerated. If a two-way command is issued to a remote device for data retrieval, the total time allotted for this transaction is approximately equal to the length of transmission time needed for the outgoing message plus the transmission time of the incoming message. If the inbound response is rejected because an error is detected in the message and a retry gives a correct response, then the previous message is probably contaminated by noise. However, if there is no response to a query, this could be caused by several other possibilities:

- the outbound command is never actually sent;
- the outbound command is contaminated by noise;
- the remote device is damaged, disconnected, power is out at the device or there is an outage in the network;
- the physical path for communication is temporarily out of service or rerouting is not available;
- a communication node is not functioning;
- the inbound transmitter at the transponder does not send the inbound message;
- the inbound receiver is not functional.

Without any communication network monitoring capability, the operator of the system will have no idea as to what causes a 'no response' to an outbound command that he thinks was issued by the server. If the system is designed to detect any of the possible causes mentioned above, the number of wasteful retries can be drastically reduced. Also the likelihood that data are lost forever during transit from a remote device to the receiver is minimized.

There are several possible causes for the degradation of communication performance. The most common is a high level of noise which contaminates the transmitted data. Applying good filtering techniques, increasing signal strength and applying error detection and correction coding, in addition to using communication retries, can dramatically improve the communication performance. Some of the error detection codes used are, for example, the Hamming error detection codes and the cyclic redundancy coding (CRC) methods. The more sophisticated error detection and correction codes are, for example, the BCH codes, the erasure filling techniques, etc. All these codes add to the length of the message and considerations such as economy of time usage for communication play a major role when large volumes of messages and data have to be transmitted.

For a quick assessment of the communication performance, the following method can be used. Assume that there are N_0 perfect transponders. Assume that outbound first try communication performance is $\eta_{out}\%$. The number of transponders that receives the outbound message is $N_0 \times (\eta_{out}/100)$. Assume an inbound first try communication performance is $\eta_{inb}\%$. The total successful two-way communication based on single try performance is $N_0 \times (\eta_{inb}/100) \times (\eta_{out}/100)\%$. Hence, the two-way communication first try performance is

$$\eta_{\text{two-way}}\% = \left[N_0 \times \left(\eta_{inb}/100\right) \times \left(\eta_{out}/100\right) \times 100\%\right]/N_0 = \left[\eta_{inb} \times \eta_{out}/100\right]\%.$$

The performance numbers can be obtained by running a pilot system specifically set up for communication testing for a sufficient period of time. The number η_{out} can be determined if the transponder is provided with an outbound communication counter. Every time an outbound message is successfully received, the counter is incremented by one. Upon retrieval of this number, it can be compared to the number of outbound commands issued by the server and η_{out} can be calculated. Once $\eta_{two\text{-}way}$ has been determined after a certain trial period, then the magnitude of η_{inb} can be obtained by dividing $\eta_{two\text{-}way}$ by η_{out}.

Some future utility applications of communication technologies play a synergistic role in improving the reliability of the infrastructure. As an example for power line communication technologies, the communication network is also the electric distribution network. Hence, the outage management function is also the communication network monitoring function. Many utilities have converted their one-line electric network drawings into digital form. Much of that information does not have the geographical information that an area map and facilities management (AM/FM) can provide. An outage management system, coupled with an AM/FM system, can improve maintenance and repair services by reducing the time for patrolling the line and by quickly identifying which protective device has operated. The information generated can also be used to determine which communication nodes are de-energized. Changes also occur if segments of the distribution network are transferred to other sources. Non-power-line based communication technology systems, such as RF networks, may or may not be affected by an outage if the RF nodes are equipped with standby power. However one still needs to be able to determine which parts of the network are out of power to avoid communication attempts to the de-energized remote devices and avoid wasteful time-consuming attempts to determine the extent of an outage. These systems are less affected by reconfiguration of the utility network.

Communication network monitoring and control and outage management are not only useful functions, but they are necessary to provide service continuity and reliability of services. When these functions are properly designed and used, unnecessary operations and delays can be avoided.

2.3 Data warehousing

Collected data which are stored are useful when they can be cross-correlated with time and physical location. Energy metering interval data which are time-synchronized can be used to determine the coincident demand of the total system. If other parameters are tagged to the metering data, such as the phase, feeder number and the substation bus, one can use this information to implement load balancing. If the location of the loads and the voltage on the feeder are added to the collected data, the information can be used to improve control of the voltage profile on the feeder through optimal scheduling of capacitor bank switching.

Synchronized, time-stamped interval metering data opens the door to numerous applications. Metering includes all energy related data such as voltage, power quality, etc. The duration of an interval depends on the application of the collected information.

For a total energy-metering device such as a kWh meter, if the reading at T_1 is P_1 and at T_2 the reading is P_2, then the magnitude of $(P_2 - P_1)$ is the energy consumed during the interval Δ, where $\Delta = (T_2 - T_1)$. The smaller the interval Δ is chosen to be, the more accurately one can characterize the load behavior. For load survey and demand metering, the value of Δ is typically 15 min. For advanced metering applications Δ is typically 30 or 60 min.

If a voltage V_1 is read at time T_1, then this is the rms voltage at T_1. The type of information one needs about the behavior of the voltage will determine what is collected, processed and stored within the time interval $\Delta = (T_2 - T_1)$. For voltage profiling of a feeder, the average value of the voltage for the time interval Δ and time stamped when the measurement was performed, is more important than short time excursions of the peak values of the voltage within the time interval Δ. Time stamping the retrieved data is essential for correlating the gathered data with other events that may occur in the distribution network. The hourly time-stamped readings provide sufficient accuracy to apportion energy usage billing by different retailers. A data warehouse can store all the time-stamped interval metered data, which can be made available to different parties.

Hourly time-stamped meter reading data which are time-synchronized can be used for other applications. The cumulative customer time interval energy usage profiles can be used to determine coincident peak demand, and the shoulders and valleys for different classes of customers, grouped by distribution substation, by distribution transformer, etc. More accurate load factor prediction for specific areas, coincident hourly load versus time, etc, can be used to develop comprehensive rate structures, for developing load control strategies and also to provide design data for future network expansion. The concept of time-stamping and synchronizing the hourly reading interval of all the meters requires a two-way communication capability that can accommodate high data throughput. Data throughput will be quite large and the communication time to obtain the data from all the remote metering transponders should not cause loss of synchronization and loss of data. Synchronized, time-stamped interval metering data and identifying the grid parameters of the location where the data are collected open the door for smart grid applications.

Knowing the physical locations of the transponders in the distribution network and their geographical locations will also greatly help the maintenance and dispatch centers improve the operational reliability of the system. Hence, for completeness, the data monitored and collected by each transponder should include a set of parameters such as the time the data are taken, the electrical network or distribution grid parameters, and geographical coordinates. Suppose the physical data that are measured are defined by the letter Q. Then it is possible to describe Q as follows:

$$Q(t, s, f, \Phi, l_x, g, \ldots),$$

where t is time, s is substation name, f is feeder number, Φ is feeder phase, l_x is feeder segment number and g is geographical location.

It is obvious why all these attributes tied to a data point are needed for comprehensive implementation of many of the utility functions. If the data collected from the remote points in the network and SCADA at the distribution substations (such as kWh, kVAh, voltage, etc) are time-stamped and synchronously monitored, and also if they can be spatially correlated to the AM/FM system, they can open the door to numerous applications that will benefit the electric utility company. For time synchronization, the National Bureau of Standards cesium clock in Boulder, Colorado, can be used in the USA. In other countries, geographic information system (GIS) clocks can be used for synchronization.

2.4 Organization and addressing methods

With such a large number of transponders from which to retrieve data, in order to control groups of units or single units scattered over a large area, the tight organization of schedules, data flow control, actions, etc, is very important. Organizing certain transponders into functional groups, which perform the same type of task and can be reached by the same communication path using one outbound group command, will simplify the communication to these transponders. This requirement implies that one has to be able to determine ahead of time or during transponder installation that they are accessible through the same communication path and should not change in the course of time. If the communication path changes, then the change has to be identified quickly. A rerouting algorithm determines the new path. Transponders, intermediate nodes and main communication nodes require power to maintain their operational capabilities. Any outages in the electrical distribution network will affect the integrity of the communication network. The ability to identify the impact of an electric outage or distribution network configuration changes on the communication infrastructure is not a luxury, but is actually a basic requirement. Avoiding wasted communication efforts to de-energized transponders, nodes, etc, or not knowing where to send the maintenance and repair crew, etc, have to be considered in the communication network design.

The central control computer has to know how to reach a transponder and where the response can be detected and received. This *path* information is stored in the database as part of the total information pertaining to that transponder. Transponders which have the same physical path can be grouped and the group address downloaded at the transponders. Single group commands can be used to address such groups. Since they communicate using the same physical path, the location where the inbound responses can be received is known ahead of time. Meter reading transponders are excellent candidates for such functional groupings. Transactions can be scheduled, queued and the group responses, after they are received and decoded, can subsequently be directed to the appropriate customer service computer. Two-way load control transponders can also be assigned into functional group addresses, although there is no immediate need for inbound group responses.

There is also a need for communication retry commands. They are needed to complement group commands in case some transponder responses from the same group are rejected because of errors in the inbound message. The retry is used on single transponders to retrieve transponder data which have previously been impaired by noise during transmission.

Based on the discussion above it is clear how the communication system is going to be used. There will be a large number of scheduled operations, there will be numerous unscheduled single operations and there will be emergency cases where other operations have to be aborted in favor of a sequence of single operations in response to an emergency alert. Hence, a method has to be designed to allow some emergency conditions to receive priority access to the system and, if necessary, to override any prior set schedules without causing loss of data and allowing error free resumption of scheduled operation after a priority interruption. It is important to understand how the communication system is going to be used and shared by various parties with different interests and priorities, as well as how to monitor and access the system.

2.5 Data security, system recovery and loss of data

Another problem is the contamination of data during transmission. This is normally due to noise. For binary encoded transmission data, techniques are available to minimize the hazard of data contamination. For error detection, CRC-16 coding is quite often used. To correct some errors in a message, the Hamming code and/or the BCH code or other error detection and correction codes are sometimes used. Any of these codes for error control are appended to the transmitted data and will increase the length of a message.

Loss of power to transponders can occur. The loss can be of short duration due to a re-closing breaker operation. When a prolonged outage occurs, the loss of power can last for hours or days. Some of the information that was downloaded to the transponder from the central computer, such as group addresses, schedules, real-time clocks and monitored data, etc, will be lost unless provisions are available at the transponder to store all the pertinent information during loss of power. Not knowing which transponders have lost their remotely downloaded addresses spells disaster. Capabilities such as ‘power ride through’ are intended to protect the system from loss of information during short duration loss of power. ‘Non-volatile’ memory devices are usually used for long duration power outages.

The consequences of the loss of a group address can be disastrous. One normally does not know when this happens and which transponder is suffering from this loss. If during a group response, one or several transponders from the group are not responding, it is impossible to know ahead of time what happened to the transponders. One might guess that there is a communication problem, the transponders are bad, etc. If ultimately one guesses correctly after several retries to communicate, then the group address has to be reloaded to each of the transponders. Without ‘non-volatile’ memory, time-stamped collected data will also be lost forever. Some of these same types of problems can also occur with data concentrators.

IOP Publishing

New Technologies for Smart Grid Operation

Sioe T Mak

Chapter 3

Reliability principles

3.1 Introduction

The *Webster's New Collegiate Dictionary* defines reliability as 'the quality or state of being reliable' and it further defines reliable as 'suitable or fit to be relied on'. It is obvious that such a definition of reliability cannot be used for engineering purposes. Unless some measure is introduced, which is accepted by both vendors and users, the meaning of reliability can be the cause of a lot of confusion, bad feelings and costly expenditure for both parties. A classic example are the antipodal meanings attached to the word reliability. From a user's standpoint any piece of equipment should perform at 100% at all times under any circumstances during the user's lifetime. This is an impossible requirement imposed on the vendor. Any conscientious vendor will do their utmost to guarantee an acceptable performance level for a certain lifetime in the most economical fashion. To provide a common base that is understandable and acceptable to both parties is not easy and ultimately the reality of the physical world and our understanding of it will be the dictating factors.

The advent of fast computing devices, miniature intelligent devices and communication technologies makes remote meter reading and smart grid implementation possible. Functions such as advanced metering infrastructure (AMI) and demand response strategic success can be measured by meeting certain performance indices. Issues such as reliability, availability and cost are the prime factors to consider in order to achieve the required performance indices. One can intuitively sense the relationship between reliability and availability which will be described later in this chapter.

To put everything into proper perspective, consider a simple hypothetical system as an example. In figure 3.1, a central control unit monitors and manages $N = 3$ remote units which are dispersed over a large geographical area. Three communication control units maintain the communications between the central control unit and the remote transponders. L_1, L_2 and L_3 link the communication control units to the central control unit. It is obvious that a failure at the central control unit is

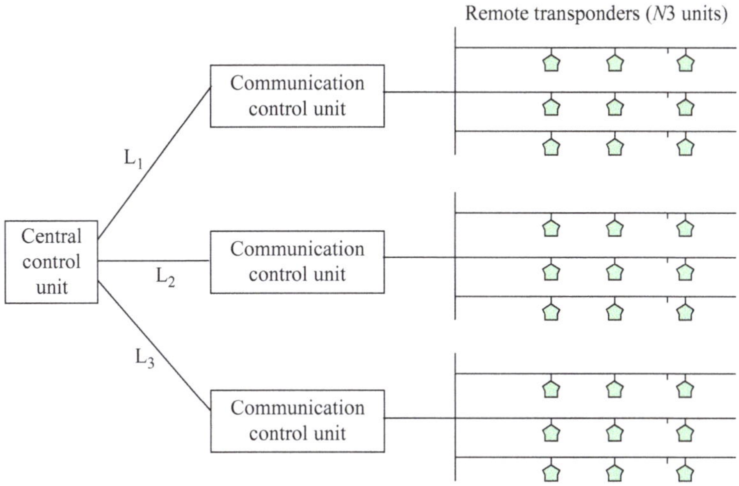

Figure 3.1. Communication system architecture to control remote intelligent devices.

disastrous for the operation of the whole system. However, good maintenance practices, the use of back-up systems and providing a protective environment to house the central control unit are expected to maintain a high degree of operational availability for the central control unit.

The impact of a failure at the communication control unit affects a large number of remote transponders. Replacement or temporary substitution is still manageable and maintenance schedules can be enforced to achieve a high degree of reliability. The environment is harsher but some degree of control is possible. At the remote sites conditions are extremely complex and unpredictable, for example:

- dispersion of units over large geographical areas;
- a very large number of units;
- extreme atmospheric environments;
- extreme electrical environments;
- extreme human environments, transponder abuse, tampering;
- communication link reliability.

A certain failure rate is tolerable. The total performance depends on the reliability of the equipment and the communication between component parts which are each of equal importance.

3.2 The bathtub curve

In focusing the issue on equipment reliability, one may ask the following questions: how does equipment fail and how is reliability defined? The following formal definition can be used: ‘reliability is the characteristic of an item expressed by the probability that it will perform a required function under stated conditions for a stated period of time’.

The words probability, condition and time are key. What constitutes a smart meter or a remote sensing device? They consist of a multitude of components which

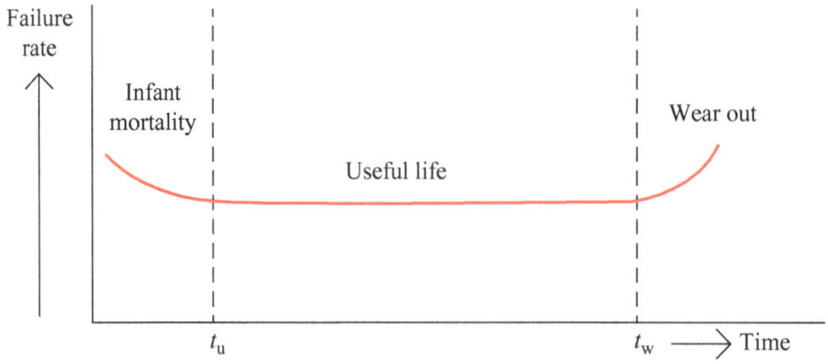

Figure 3.2. Bathtub curve depicting mean-time-between-failure (MTBF).

are interconnected and mounted on a physical supporting system inside an enclosure with links to the outside world. The software is embedded within this complex system. Each part has a probability of failure. For the whole device, one has to determine the aggregate probability of failure. Hence the reliability of one component part affects the reliability of the whole device.

The modes of failure can be catastrophic, degradation or intermittent. From a time standpoint these can be separated in terms of early failures or *infant mortality* followed by stress related failures and wear out failures or failures due to aging. The classical *bathtub curve* as shown in figure 3.2, which plots failure rate against operating time, incorporates the early failures, stress induced failures and wear out failures. This bathtub curve represents complex devices consisting of multiple components or subsystems fairly well. It is true that some components do not fit the curve. It is also true that software failures do not exhibit this type of curve, although software failure may induce physical failures.

Many factors cause early failures. Some examples are:

- faulty workmanship;
- inadequate manufacturing control;
- damage due to handling and transportation;
- inferior materials;
- errors in testing and installation.

The effects exhibit themselves as depositions of impurities, cracks, uneven protective coatings, defective hermetic seals, poor bonding and solder joints, cuts in conductive traces, burned spots due to test errors, etc.

Prior to shipping, a weeding out of problem units is performed by the vendor using burn-in tests. The weeding or screening test must be designed adequately to bring the device into the region of useful life. From a user's standpoint the region of useful life is the most important aspect. The failure rate is relatively constant and figure of merit such as *mean-time-between-failure* (MTBF) is a measure of failure rate during the useful life period. At this point it is appropriate to introduce some of the concepts from probability theory used in the theory of reliability.

Suppose that N devices of the same type, all of them working initially, are subjected to some stated condition. After a time lapse t, the number of failed units is

$N_f(t)$ and the number of survivors is $N_s(t)$. The fraction $[N_f(t)/N]$ is a measure of the reliability of the type of device.

$R(t) = \left[N_s(t)/N\right]$ is defined as the *reliability function.*

$Q(t) = \left[N_f(t)/N\right]$ is defined as the *cumulative failure function.*

Following the definitions above,

$$R(t) = 1 - Q(t). \tag{3.1}$$

The rate of failure at time t is

$$\frac{dN_f(t)}{dt} = N\frac{dQ(t)}{dt} = -N\frac{dR(t)}{dt}. \tag{3.2}$$

At time t only $N_s(t)$ survive. Since the hazard applies only to survivors, it is logical to define the hazard rate $\lambda(t)$ as follows,

$$\lambda = \frac{\text{rate of failure at time } t}{\text{survivors at time } t} = \frac{\frac{dN_f(t)}{dt}}{N_s(t)} = \frac{-N\frac{dR(t)}{dt}}{N_s(t)} = -\frac{\frac{dR(t)}{dt}}{\frac{N_s(t)}{N}} = -\frac{1}{R(t)}\frac{dR(t)}{dt}$$

or

$$\lambda(t)R(t) = -\frac{dR(t)}{dt} = \frac{dQ(t)}{dt}. \tag{3.3}$$

A quick glance at (3.3) tells us that $\lambda(t)$ has the dimension of inverse time. Since $\lambda(t)$ acts as a multiplier to $R(t)$, which is the reliability function, it is understandable why $\lambda(t)$ is called the hazard function because the product $\lambda(t)R(t)$ determines the failure rate. This product is often called *the failure density function.* To put these results in proper perspective, let us imagine the following scenario.

A batch of N initially working devices is subjected to a stated condition and the value of $\lambda(t)$ can be determined. If the same experiment is repeated n times with new batches of N each time under the same stated condition, n different numbers of $\lambda(t)$ will be obtained or, in other words, the obtained results will exhibit a statistical spread. When the same experiment is repeated, but subjected to a different stated condition, a different statistical spread is obtained. For each stated condition the median of $\lambda(t)$ is determined and defined as λ for that stated condition. The larger N and n are the more accurately the future behavior of the devices can be projected. Fortunately past experience has indicated that several probability models, together with some form of accelerated testing, will allow us to create predictor models of sufficient accuracy.

The design for accuracy can start as early as at the schematic level. The parts list, board layout and enclosure reflect the first step towards achieving a high degree of reliability for use in environments usually specified beforehand. For each part, section or block of the total equipment a reliability figure can be determined or estimated for a stated condition. The equipment reliability is a composite of the reliabilities of each part, section or block. For complex systems incorporating a large number of parts, a simple model is quite often used. For a single stated condition the regime of useful life of the bathtub curve is assumed to be flat or $\lambda(t) = \lambda$. Mathematically it is easy to show from (3.3) that $R(t) = \varepsilon^{-\lambda t}$. If each part is assigned a reliability function $R_j(t) = \varepsilon^{-\lambda_j t}$ and is assumed to be independent from the other parts, then the composite reliability function for the equipment consisting of n components is

$$R_{\text{tot}} = \prod_{j=1}^{m} \varepsilon^{-\lambda_j t} = \varepsilon^{-t\sum_{j=1}^{m}\lambda_j} = \varepsilon^{-\lambda_{\text{tot}} t} \tag{3.4}$$

$$\lambda_{\text{tot}} = \sum_{i=1}^{n} \lambda_j. \tag{3.5}$$

The inverse of λ or $[1/\lambda_{\text{tot}}]$ is called the time between failures (MTBF). The *Military Standardization Handbook* MIL-HDBK-217C uses this method to determine reliability. λ_j is a statistical entity which has to be determined from testing. Different stated conditions results in different λ_j.

Before going any further into the validity of the reliability model, let us touch upon the subject of the stated conditions as related to remote smart meters, and sensors and switches. Direct mechanical stresses, except those due to thermal conditions, are practically non-existent on the electrical parts. Stresses due to humidity, ultraviolet light and oxidation apply more to enclosures and will not be discussed here. Voltage stresses are important considerations to be taken into account. Usually utility system voltages are constrained to variations within ±15% of the nominal. However, man-made surges or surges of atmospheric origin can have a serious impact on the reliability of the equipment. Thermal stresses have by far the greatest impact on reliability. Any material in a particular energy state is bound to fail. Failure could mean a change in chemical bonds, crystal structure, etc, in such a way that the material no longer exhibits the intended useful properties. In electronic devices the results can be, for example, electromigration resulting in metal openings or voids, dissolution of silicon into aluminum films, hermetic package cracks allowing corrosion to occur due to humidity and oxidation, etc. If more energy, such as thermal, electric or other types, is added from external sources, an acceleration of changes will occur. From the perspective of a single part it is impossible to trace exactly what section in the internal structure of the part changes first and is followed by changes in other sections, or whether these changes are dependent upon each other. The best we can do is to obtain a probabilistic model of the part without knowing the exact internal events occurring inside the part. Once a failure occurs, one can then hypothesize the events leading to the failure. Provided that enough consistent failure types can be identified, from a reliability standpoint the probability model is still the model to be used.

The currently used model which relates temperature to useful life is the *Arrhenius* equation. The *IEEE Guide for the Statistical Analysis of Thermal Life Test Data* (IEEE Std 101-1972) defines the adaptation of the Arrhenius equation as follows,

$$\log_{10}(\text{life}) = \text{constant}\left(\frac{1}{2.303}\right)\frac{E}{RT} \tag{3.6}$$

or $Y = a + bX$, where $Y = \log_{10}(y)$, a = constant, $b = E/(2.303 \times R)$, $X = 1/T$, Y = life, R = gas constant, T = absolute temperature in kelvins, and E = activation energy of the reaction assumed to be constant within the applicable temperature range.

Equation (3.6) plotted on $\log_{10}(y)$ versus $1/T$ should give a straight line. Actual tests on many of the same types of component parts at one fixed temperature show a statistical distribution of Y that looks like a bell shape distribution curve. Repeat tests at different temperatures using new parts exhibit a similar distribution of Y. A straight line can be drawn through the mean values of Y for the various temperatures. Special techniques to plot the regression line incorporating confidence bounds have been developed. The straight line obtained describes the MTBF or λ as a function of temperature. The mean life at a specific temperature for a component part is used to describe the reliability function $R(t) = e^{-\lambda t}$ and forms the basis for the determination of the MTBF of an item of equipment at a specified operating temperature.

Suppose that, in the light of what has been discussed previously, the concept of MTBF and the exponential reliability function model are acceptable. How should one interpret these results properly?

$$\lambda = \frac{\frac{dQ(t)}{dt}}{\varepsilon^{-\lambda t}} = \frac{\frac{Q(t)}{dt}}{R(t)} = \frac{\frac{dQ(t)}{dt}}{1 - Q(t)} = \frac{\frac{dN_f(t)}{dt}}{1 - N_f(t)}. \tag{3.7}$$

λ = constant simply means from (3.7) that the failure rate with respect to the remaining survivors remains the same and does not depend on how many have failed. As an illustration, let us assume an MTBF of 20 years and increment the time t by one year at a time. $R(t) = N_s(t)/N$ and also take $N = 4000$, for example. Hence the number of survivors is $N_s = 4000 \times \varepsilon^{-(t/20)}$. The results are collected for $\lambda = 0.05$ in table 3.1.

These numbers are not the results of an actual test, but are computed from the reliability model. The first year failure rate is 195 yr^{-1}. The number of survivors is 3805. From here one finds $\lambda = 195/3805 = 0.0512$. For the second year $\lambda = (381 - 195)/3619 = 0.0514$. The reason that λ is not 0.05 is because of an

Table 3.1. The number of failed units as a function of time in years.

t (years)	No. of survivors	No. of units failed	No. of units failed in % with respect to original number
0	4000	0	0.00
1	3805	195	4.88
2	3619	381	9.52
3	3443	557	13.93
4	3275	725	18.13
5	3115	885	22.12
18	1626	2374	59.34
19	1547	2453	61.21
20	1472	2528	63.21

arithmetical rounding off error. If one pro-rates the failure over a period of 20 years, we obtain an average failure of $63.12\%/20 = 3.16\%\ \text{yr}^{-1}$. Hence asking for a failure rate of $0.5\%\ \text{yr}^{-1}$ based on an MTBF of 20 years is really requiring something that is incompatible with the concept of MTBF = 20 years.

Of course one can play games by pro-rating the failure over a period of 30 years. This will bring down the average failure rate to $2.59\%\ \text{yr}^{-1}$. Realistically this number game has no meaning.

The value of λ needs adjusting if other stresses are added to the thermal stresses, such as voltage and mechanical stresses. Life curves versus voltage at a stated temperature are available but usually in the area of compact dielectrics for high voltage applications. Voltage stresses in electronic equipment in relation to life tests are applied to inputs leading to an increase of input currents, leakage currents across p–n junctions, etc, which in effect increases the thermal stresses. The military has developed π for various types of component parts as correction multipliers to λ_b, defined as the base failure rate relating to the influence of electrical and temperature stresses. For discrete semiconductors a typical expression would be

$$\lambda_p = \lambda_b \prod_{j=1}^{k} \pi_j$$

where π_1 could mean the factor relating to the environment, π_2 that relating to the number of gates or posts, π_3 that relating to quality, etc.

In many instances λ_p can be a result of additive properties, such as $\lambda_p = \prod_{j=1}^{k} \pi_j + \prod_{i=1}^{l} \pi_i$. The π are determined from tests or assumptions based on similarities of design, configuration, material or complexity. Because of the high degree of availability required for short missions in very harsh environments, extremely high reliabilities are required by the military. For military grade devices, long MTBFs achieved by de-rating and redundant design and revolutionary technology tend to be the norm. These principles applied to commercial products are not economical. At the same time the military design for reliability teaches us the disciplines for reliability design. Applied to commercial products, the military criteria tend to make the commercial products using commercial grade components look less reliable, probably by a factor of two to three.

As microprocessors are being used in almost all metering and remote sensing and control devices, great attention also has to be paid to software reliability. Many peripheral devices that were at one time driven by analog devices with inherent time delays, sensing of analog inputs and conversion of analog entities into other analog entities, now perform digitally using digital devices and digital processing. What was at one time only the reliability of analog devices has now become the reliability of both digital devices and software. It is difficult to define 'software type' stresses and effects on software life, yet at the same time we intuitively feel that a 'build up' condition can occur, causing software malfunction leading to analog device failure. Electromagnetic interference (EMI) of a certain frequency and strength can introduce certain states in the program's loop by adding or deleting bits. The elusiveness of software makes it more difficult to define software reliability. Where do we go from here?

Maybe we still have to live with the bathtub curve, although there are very strong indications that the failure density function $\lambda(t)R(t)$ for semi-conductors exhibits log-normal properties with no wear-out characteristics. Mechanical parts such as relays used for switching, LEDs, opto-isolators, etc, do show wear-out characteristics. Better models confirmed by experiments and experience have to be generated. However, in the interim period there are obvious steps that can be taken for the design of reliability. These include using state-of-the-art components with proven track records and, wherever possible, using redundancy principles for the critical sections. De-rating is also a possible alternative. Also, exercising tight control on parts inspection and manufacturing processes, and using protective devices or methods to minimize the hazards of over-voltages, EMI and other environmental stresses. Watchdogs in hardware and software should be implemented to prevent run-away conditions. Any equipment that goes out of the door should go through a screening procedure that may consist of a simple visual inspection plus a burn-in test and other tests dependent on the functional or operational use of the equipment. All of these steps are part of an ongoing process.

Equipment that is housed in substations and main offices, because they are few compared to the number of remote devices dispersed over large geographical areas, are more amenable to maintenance or repair. Functionally, extremely high reliabilities are required for this equipment. A central control unit that is not functioning makes the whole system non-operational and reliability should be looked at in terms of availability or operational readiness. A proper definition of availability is

$$\text{availability} = \frac{\text{MTBF}}{\text{MTBF} + \text{MTTR}}. \tag{3.8}$$

MTTR is the total corrective maintenance time divided by the total number of corrective maintenance actions during a given period of time. Long maintenance times with a low total number of corrective maintenance actions give large values of MTTR or low availability. By the same token, short maintenance times with a large total number of corrective maintenance actions give a high availability. A good maintenance schedule of equipment with reasonable MTBF will increase the availability. Spare parts can shorten maintenance times and lower the MTTR.

For computers, one of the more serious conditions is the sneaky failure condition. Latent circuit paths, timing errors, etc, may trigger unintended actions. These type of failures can cause long MTTR. Today, several methods of self-checking for continuous diagnostic purposes are available.

3.3 The cost of reliability

The *cost of reliability* is the cost incurred when a device does not meet the operating performance requirements. As an example, a load control device is instructed to turn off an air conditioner. However, for whatever reason—lost communication or device malfunction—the load control device does not turn off the air conditioner or a remote meter does not respond to a read command, etc. There is a cost incurred because of non-performance or 'not knowing' and this is also part of the cost of reliability. If the communication performance can be predicted and each transponder can be checked for damage on a regular basis, then the cost of 'non-performance' or 'not-knowing' can be minimized. Equipment reliability costs are different for one-way systems and two-way systems. However system reliability costs are similar for both one-way and two-way systems.

If a known malfunctioning unit can be replaced immediately then equipment reliability costs refer to non-detectable malfunctions and system reliability costs relate to things such as poor communication performance, operating errors, etc. From a one-way load control standpoint, a non-detectable malfunction for a one-way system means the following:

- Field surveys have to be conducted at regular intervals at a prorated cost of K dollars per receiver. All receivers have to be checked.
- Costs associated with non-contribution to peak demand reduction or non-meter reading of failed units.

Suppose that N units have to be checked every T years. The survey cost is NK. The cost due to non-contribution of the failed units to the intended application can be calculated as follows. Assume the cost of non-contribution per unit is P per unit time and remains constant. For a time interval T after installation, following the curve shown above in figure 3.3 the cost of non-contribution is

$$C = P\left[\Delta N_{\mathrm{f}1}(T - \Delta t) + \Delta N_{\mathrm{f}2}(T - \Delta t) + \cdots + \Delta N_{\mathrm{f}n}(T - \Delta t)\right], \tag{3.9}$$

but this is the same as the cost $= P \times$ (area under the curve) or

$$C = \int_0^T N_{\mathrm{f}}(t)\mathrm{d}t. \tag{3.10}$$

From previous considerations

$$N_{\mathrm{f}}(t) = N\left[1 - \frac{N_{\mathrm{s}}(t)}{N}\right] = N[1 - R(t)].$$

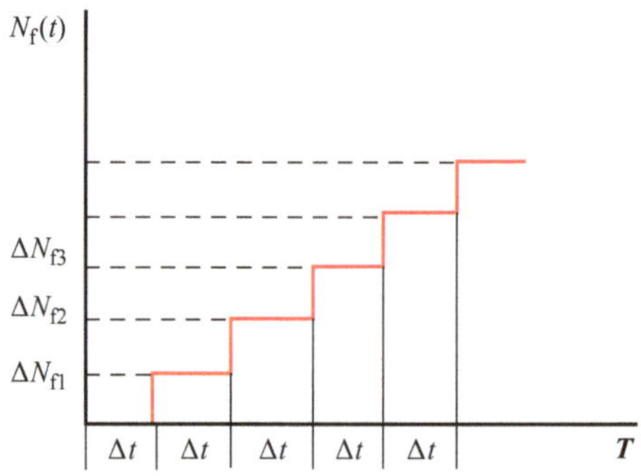

Figure 3.3. The cost of non-contribution as a function of time.

By using the useful life portion of the bathtub curve

$$N_{\mathrm{f}}(t) = N\left[1 - \mathrm{e}^{-\lambda t}\right]$$

in the equation for cost

$$C = PN\int_{0}^{T}\left(1 - \mathrm{e}^{-\lambda t}\right) = PN\left[T - \frac{1}{\lambda}\left(1 - \mathrm{e}^{-\lambda T}\right)\right], \tag{3.11}$$

the total equipment reliability cost for one-way transponders is:

$$C_T = PN\left[T - \frac{1}{\lambda}\left(1 - \mathrm{e}^{-\lambda T}\right)\right] + NK. \tag{3.12}$$

If the survey is performed once every two years, the equipment reliability cost per unit for an MTBF of 20 years would be

$$C_{T=2}\,\mathrm{unit}^{-1} = 0.09675 \times P + K$$

and on a per year basis this becomes:

$$C_{T=2}\,\mathrm{unit}^{-1}\,\mathrm{yr}^{-1} = 0.04837 \times P + K/2.$$

For a two-way system one can assume that as soon as a failed unit is detected, the unit is immediately replaced. Non-contribution to the intended application is eliminated. The reliability cost incurred is the field trip cost to replace the unit. After T years $N(T)$ units have failed. The total field trip cost will then be

$$C_T = N_{\mathrm{f}}(t) \times S$$

where S is the cost per field trip. Using the same method to calculate cost

$$C_T\,\mathrm{unit}^{-1} = \left(1 - \varepsilon^{-\frac{T}{\mathrm{MTBF}}}\right) \times S$$

and on a per year basis

$$C_T\,\mathrm{unit}^{-1}\,\mathrm{yr}^{-1} = (1/T) \times \left(1 - \varepsilon^{-\frac{T}{\mathrm{MTBF}}}\right) \times S.$$

Using the same numbers for T and MTBF as for the one-way units

$$C_{T=2}\,\mathrm{unit}^{-1}\,\mathrm{yr}^{-1} = 0.04758 \times S.$$

It should be understood that P is much larger than S and that S is closer to K. The system reliability cost is a function of the statistical properties of the communication medium, communication technology, choice of one-way or two-way and operational strategy used.

For a certain communication performance and using multiple tries, there is always a certain percentage of remote transponders that do not receive the outbound message. For one-way systems, this percentage is a statistical guess and applies only to the remaining survivors. For two-way systems this percentage can be more accurately determined and using various operational strategies it can be lowered to a known percentage. The two-way system has the advantage that the communication performance can be determined for various system conditions on a regular basis. Suppose that the reliability cost on a per year basis is denoted as C_m, for both one-way and two-way systems. The total reliability cost per unit per year can be described by the following equations

$$C_{total} = P\left[1 - \frac{\text{MTBF}}{T}\left(1 - \varepsilon^{-\frac{T}{\text{MTBF}}}\right)\right] + \frac{K}{T} + C_{m1} \text{ (for the one-way units)} \quad (3.13)$$

$$C_{total} = \frac{1}{T}\left(1 - \varepsilon^{-\frac{T}{\text{MTBF}}}\right)S + C_{m2} \text{ (for the two-way units)} \quad (3.14)$$

where P = annual cost due to lack of contribution to the intended application, T = survey period, K = cost of survey, S = cost of field trip to replace a two-way remote device, and C_m = annual cost for a certain percentage of remote devices not receiving the outbound message and therefore not contributing to the intended application. The number C_m depends on the communication technology used and its performance level. For one-way systems this number relates to survivors only.

In conclusion, everything can be summarized as follows:

- tight control has to be exercised on parts and manufacturing processes;
- part selection for a design should be based on components with proven track records;
- a burn-in test should be performed on a 100% basis on remote devices;
- MTBF and failure rates are not independent entities;
- better reliability models have to be developed based on real life data;
- upstream equipment reliability should be measured in terms of availability;
- continuous investigation is necessary to improve software reliability;
- reliability costs should receive more consideration from users.

IOP Publishing

New Technologies for Smart Grid Operation

Sioe T Mak

Chapter 4

Advanced metering and demand response applications

4.1 Customer services and demand response

The benefits of automatic meter reading (AMR) have been reported in many publications. Many utilities have been able to justify the investment in an AMR system based on the returns gained. In this section new advanced concepts which are slowly gaining acceptance in the industry are presented. Some utilities are already implementing these but they do not yet have widespread publicity.

4.1.1 Electric energy consumption metering

Interval energy metering data contain a wealth of information that benefits both customers and the electric utility itself. Hourly consumption data provide the utility with knowledge of how energy is used by individual customers as a function of time. If the interval data monitoring of all the meters is time-synchronized, the sum total of the energy consumption profiles provide the utility with the system coincident demand which can be correlated with the generation data. If the customers' energy usage profiles can be partitioned and grouped by distribution substation or by feeder or phase of a feeder, numerous new types of information can be extracted from the data. Refined control algorithms and a multitude of new applications can be designed and utilized to accomplish certain tasks. Load balancing is one example for reducing line losses.

Accurate and timely billing, the ability to retrieve metering data from inaccessible sites, fast access to single meters and timely detection of the unauthorized use of electricity have proven to benefit the electric utility industry. To improve customer relations, some utilities create websites to accommodate their customers. Customers can view their energy usage profile, how it relates to weather conditions, the effects of lifestyle, etc. Coupled to this is the opportunity to educate the customer as to how to reduce coincident peak demand, improve the efficiency of energy usage, etc.

doi:10.1088/978-0-7503-1158-8ch4

Retail wheeling is becoming a hot topic in many countries and may slowly penetrate into the USA in the near future. Through deregulation, the utilities' ownership of the generation, delivery and selling of electrical energy is broken into different companies. The energy distribution part of an electric utility forms a new company which buys bulk power from a generating company and retails the energy to the customers. Retail wheeling allows different distribution companies to sell electrical energy to customers in a competitive fashion. In some cases these companies do not necessarily own the energy delivery infrastructure or even the electric meters, but they are allowed to buy bulk energy for resale as a commodity at competitive prices. Meter ownership, communication infrastructure for meter reading, metering data collection and warehousing, and customer billing can be handled by different companies, which provide services to the energy retailers. The following example illustrates the billing scenario.

> Retailer ABC sells energy to customer A at a rate of *Y* cents per kWh. The sales contract is for a period of *X* days starting and ending at noon beginning on a certain date. For a fee the retailer ABC obtains the metering data for billing from the company, which collects and warehouses all interval time-stamped energy metering data.

Pre-pay metering has been touted for many years by several utilities in the USA and is slowly gaining ground. Pre-pay metering has been practiced in Europe for many years. A simplified pre-pay metering scenario is the following.

> A customer tells the electric utility or the energy retailer of their intention to pre-pay for a certain amount of kWh at a mutually agreed rate. The money is deposited by the customer directly to the customer service department of the utility or through a financial institution. At the end of the service contract, when the customer has used up the pre-paid amount of kWh, the customer is alerted to deposit more money for a new service agreement.

This simple illustration of a possible transaction may not be as simple as it seems from an operating standpoint. On the date and time the pre-payment agreement is in effect, the customer meter has to be read. Before the customer runs out of the pre-paid amount of kWh, the customer should be alerted to deposit more money. How soon should the customer be alerted and when the customer does not respond, should the customer be given a *period of grace*?

Another method could be to have a device attached to the kWh meter that constantly displays the amount of money left of the pre-payment amount. But this method implies the capability to download the starting time for monitoring the energy usage which is obtained from the kWh meter, convert the energy usage into a dollar value at a rate that is contractually agreed and transmit this information to the display device inside the customer's premises.

Another issue is whether the customer's service can be disconnected for non-payment after the pre-payment money is used up. Disconnecting a customer without warning may cause grave consequences to the energy provider as well as to the customer. There may be life-saving equipment at the customer's house that requires power. Business transactions through the internet system at that moment would be interrupted. Disconnecting the power also implies that a service disconnect switch is installed at the customer's premises.

Another question is how to alert the customer that it is time to deposit more money if the end of service occurs during a holiday when all financial institutions or the utilities' customer service department are closed. What provisions are then available for the customer to avoid their service being disconnected?

Another important aspect regarding pre-payment is customer education. Customers have to learn what a kWh means and what you can do with it. To help the prospective customer, his or her normal energy utilization profile can be used as an indicator of his or her expected energy usage. This will allow the customer to make intelligent decisions on how much energy to buy. Customer profiling will also help the energy retailer know how soon to alert the customer to deposit more money. These questions have to be addressed ahead of time to determine to what extent an AMR system and its communication capability can support a pre-payment application.

Remote service disconnect devices are now available and some smart meter vendors have integrated these switches with their smart meters. The existing utility AMI communication system is used to disconnect or to connect power to a residence. These are used when a tenant is moving out or a new tenant is moving in. This method is normally coupled with reading the electric meter prior to connecting or disconnecting the power to the premises. Typical applications are in areas with high turnover rates, such as college campuses, military bases, time-share apartments, etc.

Some thought has been given to using the service disconnect switch to disconnect power to customers who do not pay their electric bills or to pre-paying customers whose total energy use has reached the total of the prepaid amount. In many states, total disconnect of power to a customer, for whatever reason, is prohibited during certain periods of the year. This is true if severe hardship is imposed on the customer as penalty for non-payment of the electric bill.

Some of the basic considerations regarding these disconnecting switches are:

- contact and thermal ratings;
- the short circuit handling capability;
- switch status indicators;
- the manual operation capability;
- the device address and its link to the metering transponder address;
- on-site operational testing and activation;
- some type of customer alert before total power disconnect.

4.1.2 Gas and water metering

Gas and water metering are quite often coupled to electric meter reading transponders through short hop radio links or hard-wired links. The transmitters in the gas and water metering devices require their own power supplies. For hard-wired communication links to gas meters, special precautions are required to ensure safety at the interface. Electric safety barriers are now available.

Interval gas and water meter reading has yet to take off in a meaningful fashion. Some attempts to monitor the water usage profiles of customers by introducing interval meter reading already exist with the intent of penalizing those customers who water their lawns during times of drought. Most of the current applications are for billing only.

Quite often several gas or water meters are coupled to a particular customer's electric meter through a radio frequency (RF) link. The RF transmitter at the water or gas meter has its own battery power supply. The electric meter hence serves as a data concentrator for the gas or water meter. Water meters sometimes pose problems for installing a communication link to them. In many areas where winters can be severe, water meters are mounted inside manholes. RF signals have difficulty getting access to or from manholes with metal covers.

One very important issue to consider is the need to maintain a database of the batteries that provide power to the RF transmitters. The average operating life has to be established first, such as the date when they are installed, the operating mode of the transmitters and the shelf-life of the batteries. If the RF modules constantly transmit metering information from the water or gas meter to the data concentrator at the electric meter, this mode of operation obviously reduces the battery life faster than when the transmitters only transmit data when queried by the data concentrator. When the RF modules are not transmitting, their power consumption is merely for keeping the electronics alive and to read and store the metering data. When not in use, batteries have a shelf-life. Usually, expiration of life data are available and can be provided by the manufacturer. Hence, once the required battery information is available in the database for each RF module, then a battery replacement schedule can be set up.

4.2 Load management and demand response

In the early years of electric utilities, *load management* was introduced as a method to curtail peak demand in order to avoid the need to invest in a peaking generation unit simply to accommodate a short duration peak demand. The culprits that cause the system demand to peak are normally appliances which operate automatically in a cyclic fashion, such as water heaters and air conditioners. These appliances operate in a random cyclic fashion at random times. This load diversity usually generates an average total load that is less than the sum of individual peak demands. Under certain seasonal conditions the average total load and the peaks do not pose any problems for generation. Severely hot and long summers or cold winters may cause peak demand to exceed the normal system peak.

Several communication technologies are already in place to support load control switching in customer homes. The problem the utility faces is to determine which customers are possible candidates for participating in the load management program. Customers whose loads are flat most of the time are definitely not suitable candidates for the load management program. To develop the control algorithm needed to accomplish the demand peak shaving task, load survey data on sample loads at some customer sites are gathered and statistically processed and compared with the generation loading profile data and also assessed as to whether the coincident peak demands occur at the same time as the system peak demand. Only specific classes of customers who qualify for load management are surveyed. The KALMAN filtering technique is quite often used to determine whether homes which appear similar are expected to behave similarly to those for which the energy usage profiles are known.

A load cycling switching strategy is then used to accomplish the task. Load control switches are used to turn certain types of appliances on and off following a load cycling algorithm. Shifting of groups of customer loads with respect to other groups reduces the coincidence peak demand. Normally load control is accomplished by turning on and off load types that are cyclic in nature. Examples of such loads are central air conditioners, water heaters and electric heating systems. Essentially, the idea is to attempt to redistribute the occurrence of peaks in a way that minimizes their peak coincidence. The implementation of a demand response or load control strategy is based on clustering a population of similar appliances which have cyclic load patterns, such as air conditioners or water heaters, into several subgroups. The first subgroup is turned on for a period of time ΔT_1 and turned off for a period of ΔT_2. The second group uses the same on–off cycle but starts at a certain time delay ΔT_{delay} after the start of the first subgroup. The same time shift is applied to the other consecutive groups. After repeating this process several times, the appliances start to follow the control algorithm load pattern. The amount of peak load shifted depends on the number of subgroups and subgroup size and also on the choice of ΔT_1, ΔT_2 and the time delay ΔT_{delay} between consecutive groups. Currently, interval meter reading of all the customers during the application of a load control strategy can be used to provide information on how well the peak shifting algorithm works to curtail the total system coincident peak demand.

An added bonus to the system is obtained when this technique is properly applied. The distribution network system load factor is also improved. This also reduces the peak demand requirement for electricity generation. This example also implies that both loads are served by the same phase on the same circuit and the dispatch office knows that the second load can only be turned on after a time delay of T_{delay}. All of this information is required to implement the peak demand shift control in order to reduce circuit losses.

Load management can also be considered as a type of dispatched negative generation. This technique, when properly applied, will also improve the system load factor. In the case of an emergency when a low spinning reserve is expected and there might be a danger of causing system instability, a SCRAM command is used to disconnect all controllable loads. Figure 4.1 shows an example of a SCRAM command on an actual system shedding 975 MW in 60 s. The energy management dispatch center issued a global outbound command to all load control switches to shed load. Sixty seconds is the time needed for the communication system to send the load shed commands to all load control switches in the distribution network served by a multitude of distribution substations.

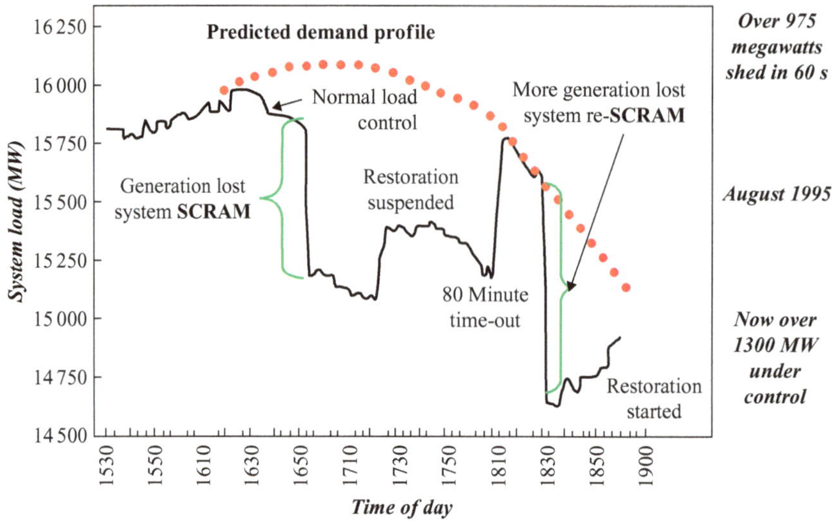

Figure 4.1. Load control SCRAM test.

After exercising the SCRAM command, the load is gradually restored by closing the load control switches in groups in a sequential order to avoid causing massive cold load pickup problems.

With advanced metering infrastructure (AMI) in place, time-synchronized interval energy metering information provides a much more accurate assessment of which customers are suitable candidates for participation in a load management program. The demand and the coincident peak demand, etc, from all customer load profiles can be obtained to design demand response strategies for reconciliation of energy usage and generation.

Energy retailers or electric distribution companies, who purchase energy from the generating companies, will in all likelihood be penalized if they allow the load to exceed a contractually agreed peak demand. Electric cooperatives in the USA understand the meaning of *demand ratchet* very well. Time-synchronized interval energy metering of all the customers served by a distribution substation can provide accurate total load and highly selective load trend data. The electric power distribution company can use this information to activate selective load control on non-essential loads and thereby prevent the total load from exceeding the demand ratchet level. The ability to produce load usage profiles for every energy user helps to categorize customers in different energy usage classes and to determine what the best control or pricing strategy is for those different classes of customer.

Another possible method takes the form of voluntary load control. Different rate structures apply for peak, shoulder and base load periods. This gave birth to the *time-of-use rate*. This type of rate structure may be seasonal, weekly or daily and may not apply to the whole population. Time-synchronized interval energy meter readings can be used to obtain a profile of how energy is used by customers and these profiles can be used to select candidates for time-of-use rates. Time-synchronized interval reading of electric meters will also generate information on how effective the time-of-use rate is in reducing total system peak demand.

Recently, home automation systems have started being advertised for smart homes. Major appliances are retrofitted with intelligent devices and linked to an intelligent master controller through a local communication network. The intelligent master controller monitors the total load and also sends switching commands to the intelligent devices in order to control the appliances, improving the total load factor and reducing the peak demand at the house. The electric utility is responsible for alerting the intelligent master controller as to when to curtail demand. A time-of-use rate could be used as the driver for how much demand to reduce. Utilities use load forecasting that may be tied to weather forecasts, the available energy for wheeling, spinning reserve requirements, etc, to determine when to activate time-of-use rates. The electric utility has to send an alarm signal to all customers on time-of-use rates to inform them that peak pricing applies during a certain period. It is then at the discretion of the customer to curtail their demand during the hours of high system peak demand.

4.3 Hybrid cars and electric vehicles

Hybrid cars and electric vehicles (EVs) are being advertised and marketed throughout the world and are purported to reduce air pollution and dependency on imported oil. However, the batteries of these cars obtain their energy from the utility network. Considering the lifestyles of the people living on the North American continent, these car batteries will most likely be charged in the evening after working hours. Typical charging times are around 8 h. Coupled with the use of appliances, such as air conditioners, cooking appliances, etc, it is expected that demand will peak during the evening hours and at night. Due to the reasons mentioned above, a secondary, much higher peak demand will occur and a new demand response technique may have to be devised to meet these new requirements. In addition, if the cost of driving an electrical vehicle is lower than the cost of driving a gasoline powered vehicle for the same number of miles, the possibility may exist that one day electric utility companies will be allowed to compete against oil companies to market their alternative form of energy for transportation. This implies that EV battery charging systems will require separate energy consumption metering.

4.4 Energy trading and/or exchange with privately owned renewable generation

As more customer owned renewable energy generation plants are installed which are not electric utility owned or connected to the utility network, new control and protection issues arise. The possibility that customer generation will produce excess energy which can be delivered to the utility network during certain times at a certain agreed price per kWh may become a new issue. Even for the case where the energy generated equals the utilities' parity level, should the electric utility pay the customer the same price as the utility charges the customer when the customer generator is not in operation?

4.5 General system alarm

When the generation spinning reserve reaches a dangerous threshold, the demand response dispatch office can issue a SCRAM command where all controllable loads are switched off. Some utilities have been considering the idea of installing alarm devices at customer premises which can be communicated with. The intent is primarily to request utility customers to reduce their load to avoid loss of system stability due to the spinning reserve margin being very low.

IOP Publishing

Chapter 5

Improvement of service reliability and optimization of energy delivery (asset management)

5.1 Introduction

Advanced metering infrastructure (AMI) and the introduction of smart meters serving as synchro-sensors generate detailed information about the behavior of the distribution network. Coupled with time information and the ability to synchronize data monitoring and save data at remote devices for retrieval by the communication system, these new technologies will further enhance the development of control functions for various applications that benefit the operation of the utility and customer services. Continuous use of this capability also provides the utility with feedback about the effectiveness of a control strategy.

5.2 SCADA serving as the master supervisor

The role of supervisory control and data acquisition (SCADA) in conjunction with AMI was initially seldom mentioned in the technical literature. SCADA systems coupled with imbedded intelligence in intelligent electronic devices (IEDs) can generate a wealth of useful information. When properly utilized and coupled with the AMI network they will allow utilities to design new applications that enhance the operation of the distribution network. SCADA becomes the master supervisor of the distribution network served by the distribution substations when coupled to the AMI communication server as shown in figure 2.1.

SCADA monitors the distribution bus voltage, the loads at all the feeders in real time and triggers an alarm when unusual conditions are detected such as faults on a feeder, overloads, etc. Smart meters serve as remote SCADA sensors whose data provide detailed information from the remote sites of the distribution network. This information, when properly synchronized in time with the SCADA operation,

doi:10.1088/978-0-7503-1158-8ch5

provides insight into the behavior of the distribution network and helps the utility design control applications to enhance the operational reliability of the utility. The SCADA system also provides feedback information on how effective a control action is in the network.

The distribution substation SCADA remote terminal unit (RTU) typically handles only the electric delivery network served by the substation bus. This gives the utility the flexibility to account for the different unique applications required for the different distribution substations.

The AMI system can play a major role in monitoring and controlling transactions and data flow between SCADA, remote devices, database management systems and the system command center. The need for stratification of control and monitoring activities by the substation bus, feeder or phase imply that remote devices can be accessed, addressed and clustered into groups or subgroups on each phase of a feeder, or on a per feeder basis for a specified distribution substation bus. Figure 2.1 shows the architecture of an AMI operational system. The AMI communication server organizes the communication addressing structure and controls all the transactions and data flow to and from their destinations, which in this case are the communication interface modules at the smart meters. The remote devices are addressable individually or by group command. This avoids communication latency due to the time needed to set up and form groups that are required for a transaction. The server is responsible for downloading the different addresses to the remote devices and also for storing them in the server database.

5.3 Dynamic network modeling of the distribution network

New tools are now available for the electric utility industry to collect more detailed information from the distribution network, which allows the operating engineer to develop more comprehensive and accurate control algorithms for optimizing energy delivery to the customers. The distribution network, including all the connected loads, is a very dynamic system from a geographical and time standpoint. The static parts of the system are the transformers (substation, step-down and distribution transformers), and the overhead and underground lines. Changes seldom occur in these parts of the network. The most dynamic parts of the network are the loads. Changes in the load distribution along the feeders, laterals, etc, as well load variations as a function of time, constitute the dynamic parts of the distribution network. Controlled changes also occur due to capacitor bank switching, voltage regulators and load control devices.

AMI in conjunction with smart meters is becoming a reality. Synchronized interval data collected from all the smart meters in the distribution network generate a lot of new and valuable information which allows utilities to develop new applications for optimizing and managing the network operation, to improve the energy delivery services to the customers and also to provide support for transmission and generation. Time-synchronized data monitoring of all loads, voltages, etc, by smart meters and the coupling of these data to circuit locations for specified intervals of time (15 min–1 h) vastly improve the outcome of many distribution functions such as

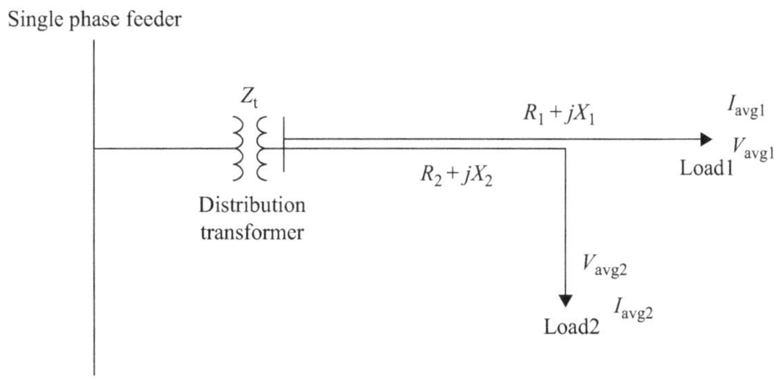

Figure 5.1. Diagram of a single phase distribution transformer serving two loads.

load balancing, integrated voltage and VAR (volt-VAR) control, recovery after a fault, etc.

Modern day smart meters can measure kWh, kVAh and the rms value of the voltage V_{rms}. Special smart meters can be used which can perform time-synchronized interval meter reading set at 15 min or longer time intervals. Average values for kWh, kVAh and the voltage V_{rms} are collected simultaneously. Hence the average power factor $\cos(\phi)$ can easily be obtained from the ratio kWh/kVAh. The average load current is equal to $I_{rms} = \text{kVAh}/V_{rms}$. Based on the extracted information, the load can be modeled as an equivalent parallel impedance R_{shunt} and jX_{shunt}.

A single-phase lateral is shown in figure 5.1 with a distribution transformer serving two loads. Smart meters at both loads can generate information such as incremental values of kWh, kVAh and the average voltage V_{avg} during a specified interval of time between T and $T + \Delta T$. The average load currents can be calculated from the information provided by the smart meters.

Assume that the incremental values ΔkWh_1, ΔkVAh_1 and the average voltage V_{AVG} are known. Then the average power factor and the average current I_{avg1} can be calculated as follows

$$\cos(\phi)_1 = \Delta\text{kWh}_1/\Delta\text{kVAh}_1 \tag{5.1}$$

$$I_{avg1} = \left(\Delta\text{kVAh}_1/\Delta T\right) V_{avg1}. \tag{5.2}$$

The load can be represented by the following parallel impedances

$$R_{shunt1} = V_{avg1}/\left(I_{avg1}\cos(\phi)_1\right) \tag{5.3}$$

$$X_{shunt1} = V_{avg1}/\left(I_{avg1}\sin(\phi)_1\right). \tag{5.4}$$

This equivalent load impedance at the secondary side of the transformer can be combined with the low voltage line impedance $(R_1 + jX_1)$ using series and parallel

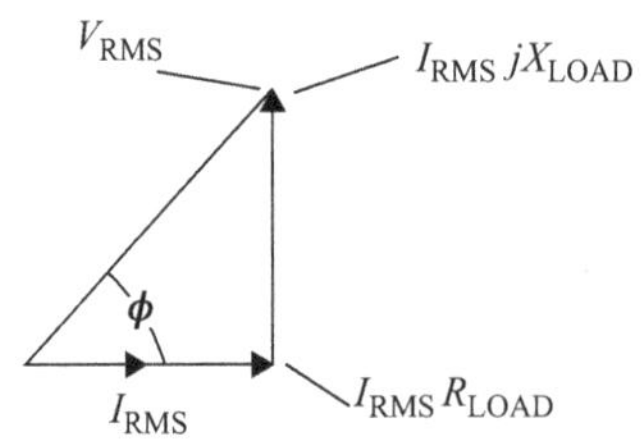

$$V_{RMS} = I_{RMS} \times [R_{LOAD} + jX_{LOAD}] = I_{RMS} \times Z \times \varepsilon^{j\phi} = I_{RMS} \times Z \times [\cos\phi + j\sin\phi]$$

$$Z = |V_{RMS}|/|V_{RMS}| = \sqrt{(R_{LOAD})^2 + (X_{LOAD})^2}$$

$$R_{LOAD} = Z\cos(\phi) \qquad X_{LOAD} = Z\sin(\phi)$$

Figure 5.2. Equivalent load impedance derived from phasor diagram.

combination methods to obtain Z_{load1}. Since I_{avg1} is known, the secondary voltage at the distribution transformer can be calculated:

$$V_{transf} = I_{avg1} \times Z_{load1}. \tag{5.5}$$

The same method can be applied to the second load for the same time interval between T and $T + \Delta T$ to obtain Z_{load2}. Both impedances can be paralleled to obtain the equivalent impedance at the service voltage side of the distribution transformer. By using the appropriate voltage ratio the equivalent impedance for both paralleled impedances Z_{load1} and Z_{load2} can be added to the distribution transformer impedance. The lateral is now reduced into one load impedance of the main feeder. The phasor diagram in figure 5.2 shows how the equivalent load impedance is obtained.

Figure 5.3 shows a three-phase feeder with a single-phase lateral serving customers using N single-phase transformers. The equivalent circuit of the subnetwork with one distribution transformer serving the two homes as depicted in figure 5.3 can easily be calculated. The equivalent load impedances are expressed in terms of a series resistance and reactance. The impedance of the first circuit from the distribution transformer to the first load is $Z_1 = (R_1 + R_{c1}) + j(X_1 + X_{c1})$. For the second circuit the equivalent circuit impedance is $Z_2 = (R_2 + R_{c2}) + j(X_2 + X_{c2})$. By paralleling Z_1 and Z_2 and adding the transformer impedance $(R_{tr} + jX_{tr})$ to the result, one obtains the equivalent impedance of the subnetwork protected by fuse F_1. This process can be repeated for all N subnetworks. The final process is to obtain the net equivalent impedance of the lateral by using series and parallel combination operations. The lateral subnetwork protected by the fuse F_{lat} is now replaced by a single impedance.

Excellent computer modeling for three-phase areal lines or underground cables is available and single-phase laterals including the loads can be connected as single taps to the feeders. The whole three-phase feeder is now replaced by a three-phase network with simple taps where each tap represents the single-phase equivalent

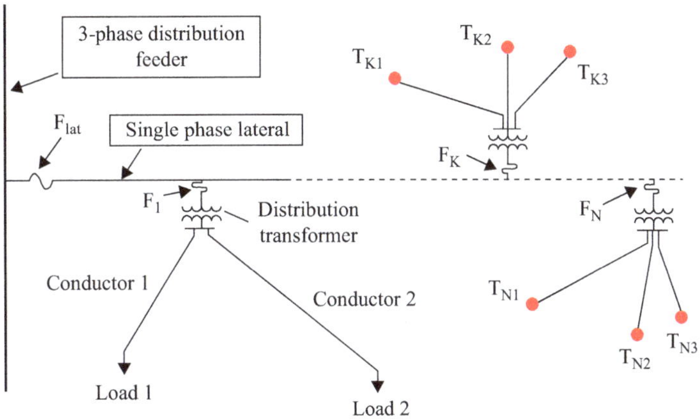

Load 1 impedance : $R_1 + jX_1$ in series with conductor 1 impedance $R_{c1} + jX_{c1}$

Load 2 impedance : $R_2 + jX_2$ in series with conductor 2 impedance $R_{c2} + jX_{c2}$

Transformer nimpedance : $R_{tr} + jX_{tr}$

Figure 5.3. Equivalent impedence of a distribution network.

impedance of the lateral subnetwork. This simplifies the complex feeder network dramatically and helps to determine through network modeling and analysis which actions to take to optimize the distribution of electric energy. Similar techniques can be applied to two-phase and three-phase laterals. The dynamic changes at the feeder are caused by the load changes. The synchronized interval data of kWh, kVAh and the rms voltage V_{rms} at the load side describe the feeder network dynamic changes with time. This type of information is useful for assessing the voltage distribution at the feeder and the losses incurred at the feeder and laterals. Control algorithms can be designed to improve the feeder load factor and the load and voltage balance, and also to construct a more comprehensive capacitor bank switching schedule. Daily, weekly and seasonal load profiles also help to adjust the controls to reflect the dynamic changes in the system.

The ability of SCADA to monitor the current of each phase of a feeder and the bus voltage serving the phase of the feeder in real time (time-synchronized with the current) and knowledge of the source impedance allows one to create a Thevenin equivalent circuit of the source. If the feeder circuit segment parameters and the distribution transformer impedances are known, then the smart meters can also provide load and other data which allow the engineer to develop single customer circuit load models. The part of the network served by the phase voltage at the substation can be constructed from the information delivered by SCADA and the smart meters. For every time interval ΔT of monitored data, the generated equivalent circuit model will be sufficiently accurate and is valid only for that time interval. A dynamic equivalent circuit model of the distribution network is obtained. Of course this assumption is only valid assuming that steady state conditions prevail throughout the time interval ΔT.

5.4 Electric utility network outage management and system restoration

5.4.1 Selective coordination of protective devices

The electric distribution network is part of the energy delivery infrastructure operating at a medium and service voltage level. The network is complex, serves a large number of customers and covers a large geographical area. To minimize the loss of continuity of energy delivery to customers due to faults in the network, protective devices are extensively used. Fast-acting isolating devices such as fuses, re-closers or circuit breakers are installed and set to operate in a hierarchical fashion. *Selective coordination* of protective devices has been practiced by many electric utility engineers. This method is one form of local intelligence which coordinates the operating time responses of the different protective devices to selectively isolate faults in the network from the remaining sound parts. The fastest operating protective devices are typically fuses or fuse cut-outs; circuit breakers and re-closers are slower and their operating times are programmable. Usually the fastest operating devices are installed at single-phase remote loads. Breakers and re-closers are usually three-phase electromechanical protective devices and are installed on the main feeder or three-phase laterals. Fast-acting fuses which typically isolate a circuit within a cycle or less are located at the more remote locations from the bus. Hence a remote fault is quickly isolated by the fuse before a main breaker, which needs several cycles to decide whether to open or not, has a chance to open.

Outages do occur in electric distribution networks. The effects of an outage are felt not only by the users of electric power, but may also affect communication infrastructure. Outages can occur at an individual customer premises, for a group of customers on a lateral served by a distribution feeder, or for a large area due to a fault on the transmission line. When an outage occurs, not only are remote communication devices out of power, but some communication nodes become de-energized as well if no provision is made for a standby power source.

Single-customer outages inside the customer home, total distribution network blackouts due to a failure in the distribution substation and failure on the transmission line side will not be discussed here. Most outages occur due to line faults at the distribution feeder or its lateral, which cause protective devices to disconnect certain segments of the feeder network. They can last for a fraction of a second due to re-closer operation, but some can last for many hours or days if damage occurs on the line or pole.

The occurrence of an outage is unpredictable and we define *outage management* as the ability to detect an outage and its extent, alert the system operator of the outage and notify the necessary departments of the event. Fault isolation and system recovery are part of the outage management function.

By properly coordinating the protective devices (selective coordination), the faulted segment of the network is isolated to the nearest fault location. Hence, all transponders connected to the part of the circuit that is de-energized will be out of power. Systems that use the distribution network for communication cannot communicate with the de-energized transponders. The same applies to other fixed network systems, unless the transponders have batteries.

The ability to monitor transponders that are de-energized, which can be related to the part of the network that is de-energized, is defined as *outage mapping*. For a properly coordinated protection system, when the fault occurs beyond a certain protective device, it will isolate the faulty section of the network from the source. This automatic isolation of part of the network is only the first step in the *fault isolation* process. The fault occurs only in a small segment of the de-energized part of the network. Due to the vast expanse of the network, it is often quite difficult to identify the location of the fault and the part of the network that is de-energized. Telephone calls from customers who have lost power and relating their addresses to geographical locations using maps have been the methods used by repair and maintenance crews to determine the most likely location of the fault.

The requirement of being able to read all the electric meters served by a utility network provides the capability to add new applications such as the popular outage mapping function. The meter reading devices—henceforth called *transponders*—in conjunction with the communication system can play an important role in the determination of the de-energized segments of an electric network.

At one time suggestions were made to apply *continuous polling* of all the transponders. This is fine if very high-speed two-way communication technology is available and the polling action does not take away the ability to perform other equally important functions. For slower communication technologies, the best method is the use of *selective polling*. If an alarm can activate the polling sequence on the parts of the network where a fault is detected, then the communication for outage mapping is only active during the polling sequence. The more precise the information content of the alarm is, the shorter the time needed for polling. The information content of the alarm can be the distribution substation bus number, the feeder and phase number, line-to-neutral or line-to-line fault, etc, which are typically available from a SCADA system.

Technologies which use low frequency signals and also use the distribution circuit as their communication medium have certain advantages that simplify the detection of faults in the network. The following sections explain in detail the aspects mentioned above.

A typical distribution network is served by a distribution substation transformer, which converts the transmission voltage to a medium voltage level. Multiple feeders at the medium voltage ranging between 4.0 kV and 35 kV deliver power to customers over a certain geographical area. Customers are served at the service voltage level through step-down transformers. A typical distribution circuit is shown in figure 5.4.

5.4.2 The propagation of fault currents in a distribution network

Assume that a short circuit occurs at the secondary of a single-phase distribution transformer. Assume that the 240 V secondary voltage phasor corresponds to the distribution substation bus voltage phasor V_{BN}. The reflected fault current I_{fault} can be detected at the phase B bus or feeder current transformer. If the fault current causes a protective device at the distribution transformer to operate, then all customers served by that distribution transformer experience loss of power. Hence, if a fault current

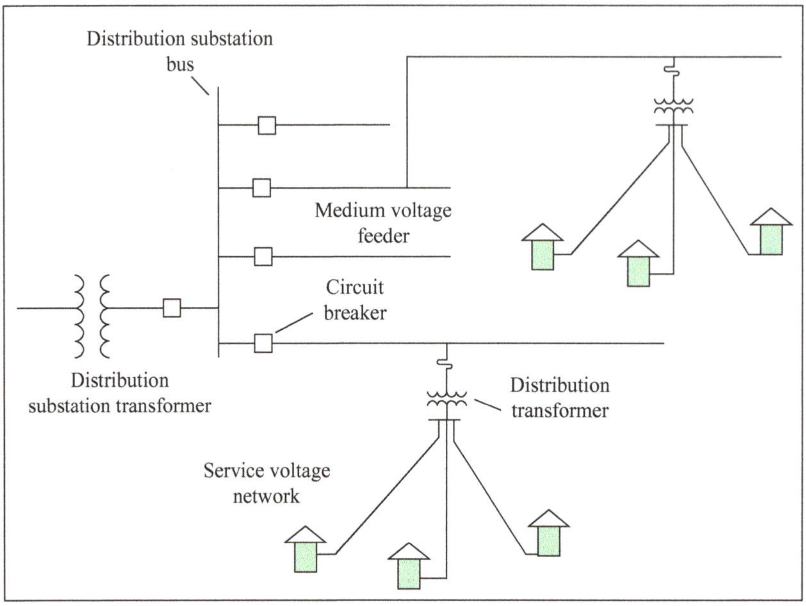

Figure 5.4. A typical distribution network.

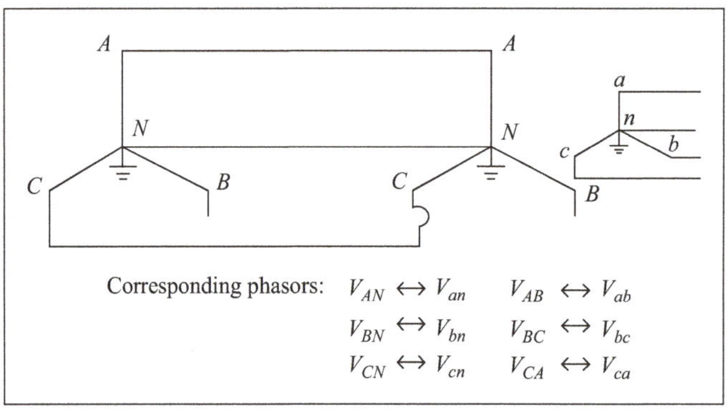

Figure 5.5. The corresponding phasors diagram for Grd WYE-Grd WYE stepdown transformer.

is sensed at the phase B current transformer, one can infer that somewhere in the network the fault has occurred at a phase which corresponds to the voltage V_{BN} at the substation.

Simple examples are depicted in figures 5.5 and 5.6. A Grd Y distribution substation transformer serves a Grd Y–Grd Y step-down transformer in figure 5.5. In figure 5.6 the step-down transformer is connected in Delta–Grd Y. The corresponding phasors are shown in the figures. For the Delta–Grd Y arrangement the phasor V_{an} corresponds to the phasor V_{AB} at the substation. A short circuit at phase V_{an} can be sensed at the current transformers on phases A and B at the distribution substation. This network behavior leads to one important application. If there are devices that can be remotely polled by a communication system, only those units on

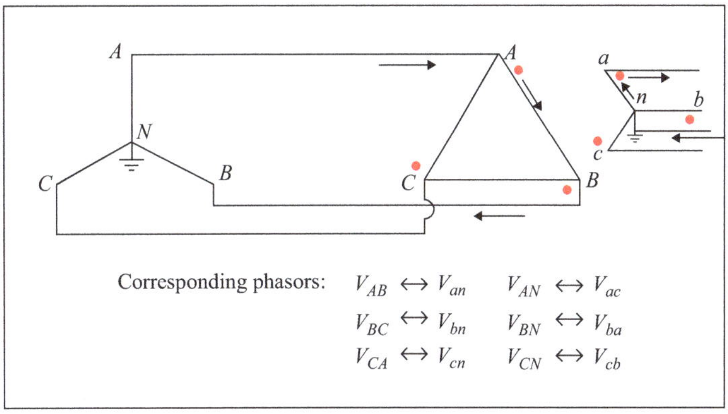

Figure 5.6. Corresponding phasor diagram for DELTA-Grd WYE stepdown transformer.

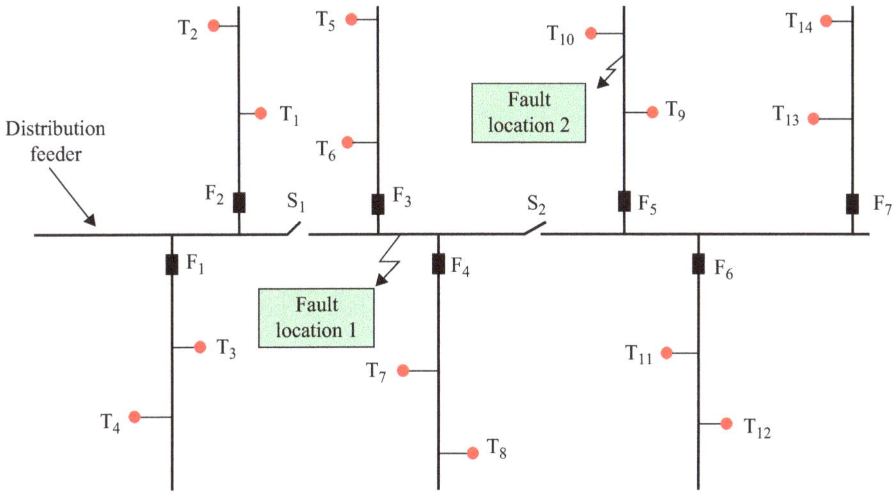

Figure 5.7. Selective coordination of protective devices to minimize effects of faults at a distribution network.

the phase and feeder where the fault is sensed at the distribution substation feeder have to be polled.

The intelligent transreceiver is located at the distribution substation. It handles all communication transactions between the substation and the transponders and also provides the high-speed link to the server. This coincidence between the fault path and the low frequency power-line communication path simplifies fault detection in the distribution network using the low frequency two-way communication system and hence the low frequency power-line two-way communication system for automatic meter reading (AMR) can be used for outage mapping. A typical operating scenario is described in the following sections.

The feeders and laterals of a distribution circuit are protected using circuit breakers and fuses. These protective devices are selectively coordinated to minimize the effects of a fault. A simple illustration of selective coordination is shown in figure 5.7. A main feeder has two circuit breakers S_1 and S_2 to protect the feeder from major faults. Many laterals are tapped from the main feeder to serve distribution transformers, which subsequently step-down the voltage to the service voltage level. A fuse is used to protect a lateral. A major fault on the feeder as depicted in the figure causes a single-phase breaker S_1 to open. The two sublaterals protected by fuses F_1 and F_2 remain energized. If a fault occurs at a lateral, which is protected by fuse F_5, only fuse F_5 will operate. Circuit breakers S_1 and S_2 remain closed. Fuses typically operate within one cycle of the power frequency voltage. Circuit breakers have delayed operation lasting for many cycles to allow fuses take care of the fault on the lateral first.

5.4.3 Selective polling sequences triggered by an alarm

If a fault current sensing device is coupled to the current transformers at the beginning of each feeder, the sensing device can establish the phase and feeder where the fault occurs. The sensing device also provides a trigger alarm for the power line communication system to poll all transponders for meter reading, T_1, T_2,..., T_{13}, T_{14}. Assume that the transponders are all on phase *A* and the fault is line-to-neutral on phase *A* at fault location 1, as indicated in figure 5.7, and the single-phase circuit breaker S_1 opens. Transponders T_1, T_2, T_3 and T_4 will respond to a polling query on phase *A*. The other transponders T_5,..., T_{14} do not respond to the polling query because they are de-energized. An inference is made to conclude that circuit breaker S_1 is open.

If a fault occurs on a lateral as indicated by fault location 2 in figure 5.7, polling of all the transponders indicates that T_9 and T_{10} do not respond to a query. Hence, by inference one concludes that fuse F_5 is open.

If the fault monitor can distinguish between a circuit breaker operation and a fuse operation, some simplification of polling can be achieved for the case when a fault is cleared by a circuit breaker operation. In the example in figure 5.7, when the fault occurs at fault location 1, one only polls transponders T_1, T_6 and T_9.

5.4.4 The intelligent fault detector

The fault detector's task is to determine that a fault has indeed occurred on a feeder. In this section, high impedance faults are excluded from the discussion. Faults can be single-phase to neutral, phase to phase and three-phase. A fault on a lateral protected by a high voltage fuse typically lasts for a few half cycles of the 60 Hz wave. A fault on the main feeder lasts for more than a few cycles and a re-closer allows several re-closing actions, depending upon the coordination scheme used on the feeder. The net result of a re-closer action due to a fault can be a lockout or a total restoration of the feeder load. The magnitude of a fault current at the service voltage level beyond a distribution transformer can be of the same order of magnitude as a medium sized load increase on a feeder. Distinguishing the different types of events occurring on a distribution feeder will require some level of intelligence of the fault monitor. This intelligence is required to avoid delivering false alarm triggers for polling.

The first requirement is the ability to determine that a fault has occurred. The method used to determine what type of fault has occurred will determine the intelligence to be used for discrimination of the various types of faults. This in turn causes the fault monitor to send the correct message to the communication system to perform selective polling of the transponders.

Several techniques are available to extract the fault current from the feeder. In general the magnitude of the fault current is much larger than that of the feeder load current. By comparing the peak value of the fault current to the average peak load current prior to the fault, a decision can be made to indicate that a fault has occurred. This section describes the use of a modified inbound signal detection algorithm of the patented TWACS power frequency communication technology, which is now extensively used for AMR by many electric utilities. The differential technique is used to extract the TWACS inbound signals from the feeder load currents and the method can be described as follows.

If the ac is described by the function $F(t)$ and, at some time before, was displaced by N times the period T of the ac described by the function $F(t - N \times T)$, the consequent value of the residue is:

$$\begin{aligned}\delta(t) &= F(t) - F(t - N \times T) \\ &= 0 \text{ (for steady state conditions)} \\ &\neq 0 \text{ (for non-steady state conditions).}\end{aligned} \tag{5.6}$$

This technique is also rooted in the Fourier transform method. A change in current magnitude (a fault or change in the load current) at $(N \times T)$ seconds beyond the time t results in a residue equal to $\delta(t + N \times T)$, which is not zero. This implies that all the sampled current data between the time t and $(t + N \times T)$ have to be stored in memory for processing by the algorithm as represented by (5.6). Instead of storing all the sampled current data, one can use the peak values of the ac waveform. This method is depicted in figure 5.8.

The differential technique is applied to some simple examples for purposes of illustration. Figure 5.9 depicts a case for fuse operation followed by a loss of some load. Figure 5.10 depicts a case for breaker operation followed by total loss of load. Figure 5.11 depicts a similar case as shown in figure 5.10 but subsequently followed by a fault clearance and total load restoration. In figure 5.11 the differential technique is applied to a load increase. The algorithm uses $N \times T$ equal to 10 cycles. The upper parts of figures 5.9–5.11 show a sequence of sampled

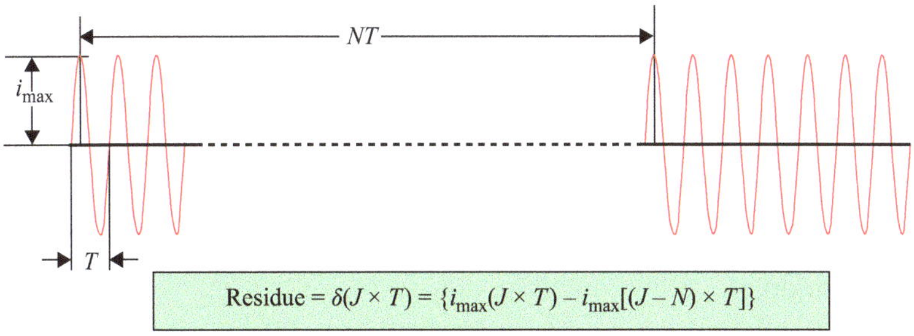

Figure 5.8. Example of the use of the differential technique.

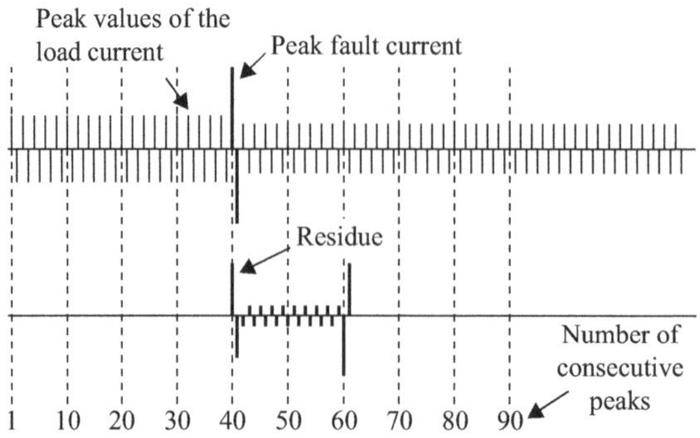

Figure 5.9. One cycle high voltage fuse operation.

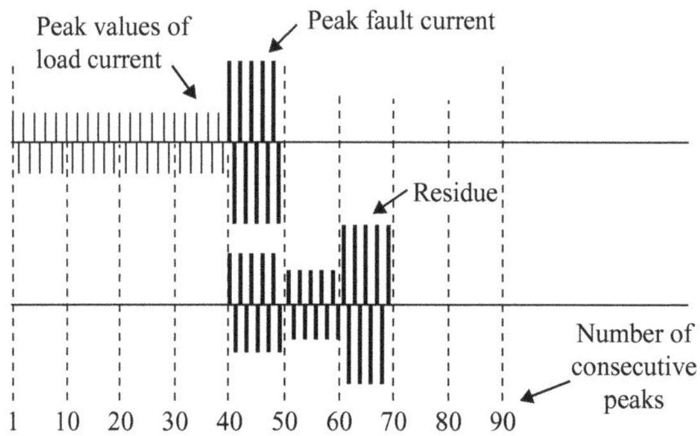

Figure 5.10. Five cycle breaker operation with load disconnect.

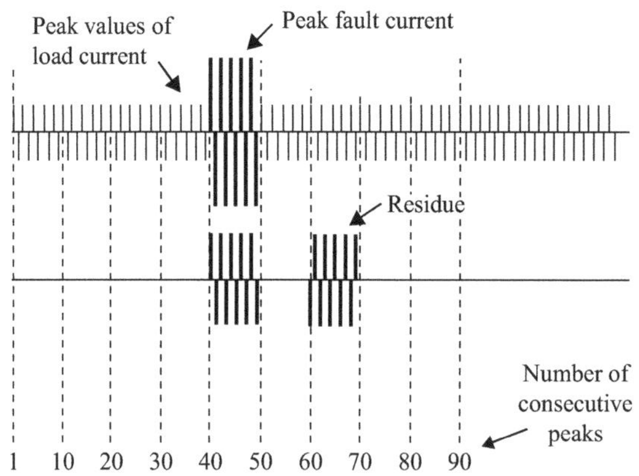

Figure 5.11. Five cycle breaker operation with no load disconnect.

current peaks. The bold portions are the fault current peaks. The residues are shown in the middle.

For the four different cases, it is interesting to note that the unique differences between the four different types of current magnitude change. In all four cases, the algorithm extracts only the main culprits that cause the changes in the current magnitudes.

The unique differences between each of the residues can be used to design an intelligent system that can distinguish between the different types of faults interrupting device operation and load changes. Simple pattern recognition methods are currently available which can compare patterns obtained by the detector with typical patterns which are computer generated and stored in memory. Very sophisticated software, such as PSCAD, which is a commercial version of the Electro-magnetic Transient Program (EMTP) can be used to generate models for circuit breaker, re-closer or fuse operation due to faults in an electric utility distribution network. The digitized data from the models can then be processed using the differential technique algorithm to generate signature patterns for reference.

Comparing the different residue patterns, it is clear that the choice of a larger value of $(N \times T)$ makes it easier to differentiate the various types of protective devices used.

5.4.5 The granularity of detection, auxiliary requirements and speed of response

The magnitude of most fault currents due to faults on the feeders or laterals are easily distinguishable from a load current increase due to a large load increment on the feeder. This may not be true for trying to discriminate the fault current at the service voltage side from a load increase current from a magnitude standpoint. To obtain a high granularity of fault detection capability, the pattern recognition capability algorithm for differentiating a load increase caused by a fault has to be made very strong. Distribution transformers are protected by fuses which typically operate in fewer than two cycles. The residue will be similar to what is shown in figure 5.9.

A solid fault at the 240 V circuit from a 25 kVA distribution transformer has a peak value between 100 A and 150 A at the 13.8 kV side of the feeder. From a load increment standpoint, these numbers mean load increments between 2.4 MVA and 3.6 MVA. Only on rare occasions does one see such a sudden increase of load on a feeder.

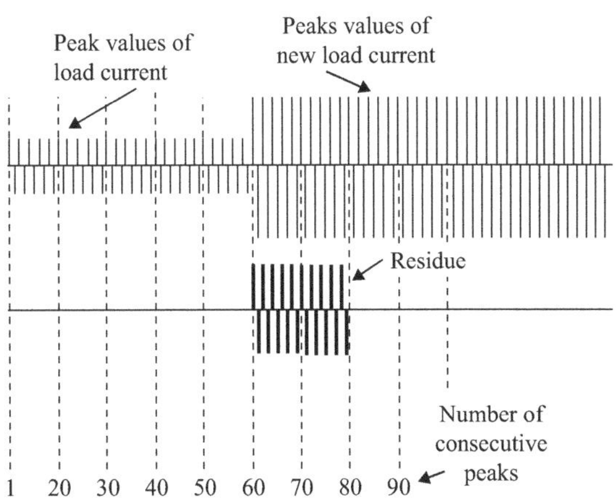

Figure 5.12. A sudden load increase.

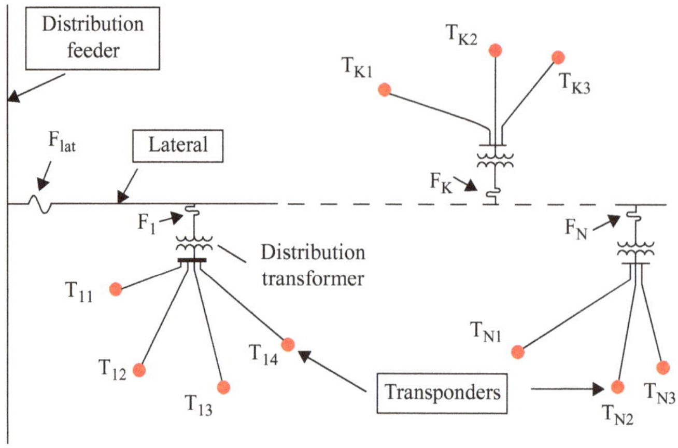

Figure 5.13. Choosing one transponder as candidates per distribution transformer for polling.

Based on the simplified patterns as depicted in figures 5.9–5.12, the number of transponders to poll is the highest for faults on laterals which are protected by fuses. Since there are no transponders directly connected to the medium voltage lateral, a fault can be at the service voltage side or directly on the medium voltage lateral conductors. In both cases the residue pattern will look like the one depicted in figure 5.9.

If there are N distribution transformers connected to a single-phase lateral as shown in figure 5.13, one can choose one transponder per distribution transformer as candidates for polling. In the example above, assume that T_{11}, T_{K1} and T_{N1} are assigned to be polled for outage mapping applications. If none of them responds to a polling query then one can infer that a fault has occurred on the medium voltage lateral conductors. The fuse F_{lat} has operated. If all the transponders respond except for T_{K1}, then the fault is either at the distribution transformer or at the service voltage side of the transformer. The fuse F_K has operated.

Since the transponders are all connected to the same phase voltage, the time needed for polling can be reduced by using the same method of group addressing in conjunction with a multi-channel inbound command. This method is identical to that already used for remote meter reading except that the units selected are strictly for outage mapping polling which is triggered by an alarm.

If n simultaneous channels are available for the inbound communication, then the first n transponders can respond simultaneously to a polling command. The second group of n transponders follows immediately after the first group has finished sending their response. Subsequently, the next group of n follows and this continues until all transponders have responded to a polling query. Only one outbound group command is required. The total polling communication time in the utility network to obtain responses from the transponders will be the time for the outbound group command T_{outb} plus the inbound response time $k \times T_{subgroup}$ plus a small time delay between the outbound command and inbound response T_{delay}. $T_{subgroup}$ is the time for one inbound response from the transponder.

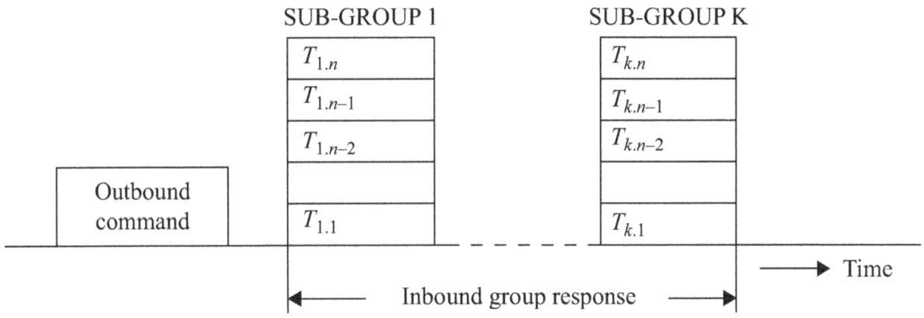

Figure 5.14. Grouping of transponders using multi-channel inbound communication as response to a single outbound command.

This group response method, as shown in figure 5.14, substantially reduces the polling time for the transponders. The additional time which has to be added is the communication time between the main transreceiver at the distribution substation and the main server.

5.4.6 Auxiliary requirements

The polling of the transponders after a fault merely identifies that some transponders do not respond to a query. The de-energized transponders are only known by their communication addresses, the phase and the feeder location. Which protective device has operated and what segment of the circuit is de-energized still have to be determined. To make this application useful for outage mapping, it is necessary to relate the transponders' communication addresses to their physical locations in the distribution network. Many electric utilities have digitized their distribution circuit diagrams and some systems even relate the circuits to their geographical locations. Area map and facilities management (AM/FM) systems are available and are already in use by many utilities. A relational database of transponder addresses for outage mapping can be established which relates the transponder communication addresses to the circuit segments and the protective devices in the AM/FM database. This synergistic relationship between the polling results and the AM/FM system also provides a real-time update of the latest state of the distribution network.

A display of the geographical location of the de-energized network segment on the dispatch computer screen will help the repair crew find the fault location. Hence, a good database of the physical locations of the protective devices, circuit segments and other circuit information related to transponder addresses used for detecting outages is necessary.

This outage information also helps to avoid unnecessary communications to the remote devices that are de-energized and the same information is also used to identify subnodes that are de-energized. Fault isolation is accomplished by opening breaker or switch S_2 and restoration is accomplished by connecting the part of the network beyond S_2 to another source. As simple as it may sound, restoration is a very complex function if one tries to automate and optimize the transfer.

If several options are available regarding where to connect the network beyond S_2, what is the basis for considering a particular option if the expected repair will take a long time? Some reasons are listed below:

- overloading of the other source after a certain time;
- difficulties adjusting voltage regulation and capacitor bank control to minimize losses while maintaining a good voltage profile.

5.5 Feeder load balancing and loss management

5.5.1 Line loss determination and control

For a specific bus, feeder and phase, SCADA can provide the total substation load, voltage and current data. If the SCADA system can be provided with interval energy reading in synchronism with the interval meter reading at the customer premises and also selectively partition the result by bus, feeder and phase, then the coincident demand due to customer load on each phase of a feeder can be correlated to the total load profile that the SCADA provides. The information can be used to determine the line loss on a phase of the feeder in question.

The total coincident customer load demand profile obtained by summing the demand profiles of each smart meter served by the phase wire of the feeder is then subtracted from the SCADA demand profile of the same phase of the feeder. The net result is the feeder loss profile for the same time interval duration of the SCADA and coincident customer demand profile. This information, which can be extracted from the database, is valuable for determining how the line losses behave. When the maximum occurs, by going through the individual customer load profiles, one can determine which of the customers has caused the line loss to maximize. If in addition the AMI database also contains information of the phase wire serving the customers, one can also identify all the culprits causing the line losses to maximize.

All the information generated above will help the utility to design strategies to reduce circuit conductor losses. Some options available are to exercise load control of controllable loads at the customer premises to improve the load factor. Another option is to transfer load to another lightly loaded feeder/phase in a seamless fashion. If a detailed study indicates that some customers can be permanently served by a different feeder without impairing the operation of the feeder, then this will be the most desirable solution.

Figure 5.15 illustrates the concept of distribution network optimization. A single-phase feeder with a line impedance $(R + jX)$ from a medium voltage substation bus supplies load currents to load 1 and load 2 for a period of time ΔT. The net load current is equal to $(I_1 + I_2)$. If the bus SCADA RTU monitors the net delivered energy for a period ΔT and all the smart meters served by the feeder synchronously monitor the consumption at each load for the same duration of time ΔT, then for scenario 1, when both loads coincide in real time, the total circuit loss at the feeder is equal to

$$P_{\text{loss}} = [(I_1 + I_2)^2 R\Delta T = \left[(I_1)^2 R + (I_2)^2 R + (2I_1 I_2)R\right] \times \Delta T. \qquad (5.7)$$

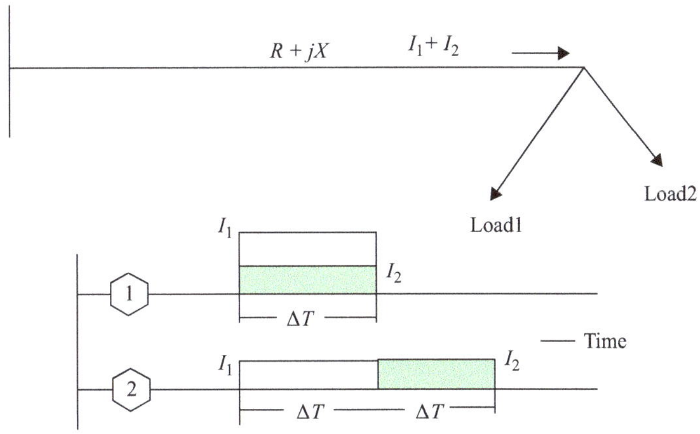

Figure 5.15. Loss reduction and voltage improvement by time shifting of load.

For the case when one assumes for simplicity that the absolute values of both currents are equal, then for $I_1 = I_2 = I$, the total loss in the line is $P_{loss} = 4I^2R \times \Delta T$. This number can be obtained by subtracting the net delivered energy as measured by the SCADA RTU for the period of time ΔT minus the sum of all the monitored energy by the smart meters for the same period of time ΔT. In other words,

$$P_{loss} = P_{SCADA} - \left(P_{load1} + P_{load2}\right). \tag{5.8}$$

Hence the SCADA time interval energy monitoring capability in conjunction with smart metering time interval energy consumption data can be used to determine the feeder losses. This is only true if the monitoring operations are synchronized.

For scenario 2 type loading, the loading interval is equal to $2\Delta T$.

The total line losses is equal to

$$P_{loss} = \left[(I_1)^2 R\right] \times \Delta T + \left[(I_2)^2 R\right] \times \Delta T. \tag{5.9}$$

If the current I_1 is equal to I_2 in magnitude and phase, the line losses for scenario 2 is equal to $P_{loss} = [2I^2R] \times \Delta T$ and is reduced by 50%. Here is a case where load shifting to reduce the coincident peak demand improves the load factor and a dramatic reduction in line losses is accomplished. In addition the peak demand on transmission and generation is also reduced. Another benefit is a reduction in voltage drop at the loads. The voltage drop in scenario 1 is higher by virtue of the larger current.

5.5.2 Feeder load balancing

Even though a three-phase source can be perfectly balanced, due to uneven load distribution amongst the phases, especially along the distribution feeders, the unequal voltage drops on the different phases cause voltage imbalances at the various locations on a feeder.

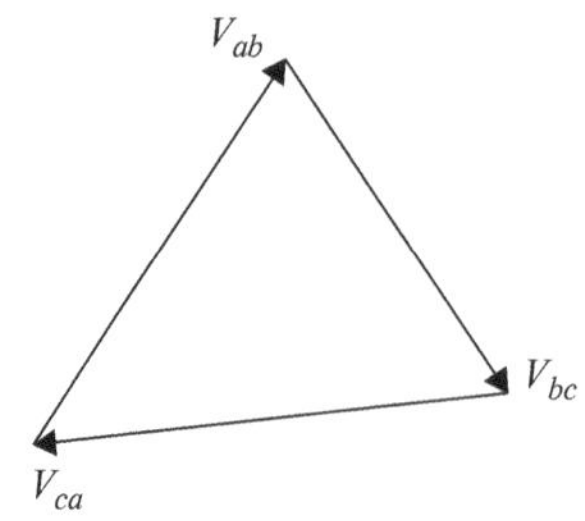

Figure 5.16. The line to line voltage triangle.

Under steady state conditions, unbalanced voltages can be decomposed into positive-, negative- and zero-sequence voltages. In three-phase electric machines, the negative-sequence voltages generate negative-sequence magnetizing currents. These negative-sequence magnetizing currents generate a rotating magnetic field, which rotates in the reverse direction with respect to the rotor. This negatively rotating magnetic field creates a retardation torque. In addition, due to the reverse rotation of the negative-sequence torque with respect to the rotating rotor, the additional induced eddy current losses by the reverse rotational field in the magnetic material of the machine will increase dramatically. Operating engineers quite often prefer to express the imbalance in terms of dissymmetry, which is defined as the ratio of the negative-sequence voltage to the positive-sequence voltage.

To express this in terms of the actual voltages, if the three-phase line-to-line voltages form a triangle as shown in figure 5.16 and the absolute voltages can be expressed as $a = |V_{ab}|$, $b = |V_{bc}|$ and the third one as $c = |V_{ca}|$ then the *voltage imbalance factor* is

$$\left|\frac{V_2}{V_1}\right| = \frac{\sqrt{\dfrac{1+x^2+y^2}{6} - \sqrt{\dfrac{(1+x+y)(x+y-1)(1+y-x)(1+x-y)}{12}}}}{\sqrt{\dfrac{1+x^2+y^2}{6} + \sqrt{\dfrac{(1+x+y)(x+y-1)(1+y-x)(1+x-y)}{12}}}}. \quad (5.10)$$

In this expression, assuming that $|V_{ab}|$ has the largest magnitude, x and y have the following values:

$$x = b/a \qquad \text{and} \qquad y = c/a$$

In many European countries the standards limit the imbalance to 2% at the medium voltage level. In the USA ANSI C84.1 Annex 1 and NEMA MG1, for voltage imbalances in excess of 1%, a de-rating of motors is required. Unbalanced loading of a distribution feeder also dramatically increases the copper losses.

In grounded wye three-phase networks (shown in figure 5.17), the sum of the phase currents in the phase wires flows in the neutral which is quite often grounded. The neutral current also quite often generates stray currents in the soil in the vicinity of the neutral wire. This is due to the standard practice of grounding the neutral wire at the poles and also at the step-down transformers. In practically all instances, the center tap in the 240 V service voltage side of the transformer is also grounded and tied to the ground at the distribution voltage side. In addition, the safety ground at

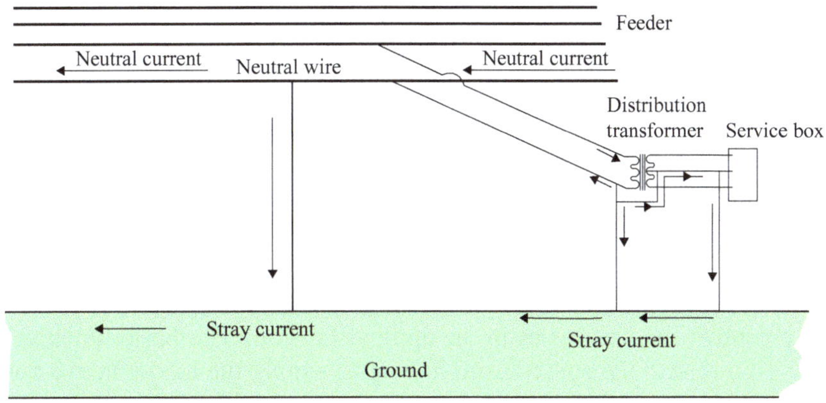

Figure 5.17. Stray currents causing stepvoltage.

the customer premises is tied to the incoming neutral wire. As a result, at each distribution transformer high side grounded neutral, other remote grounds provide additional grounding shunts. This can lead to stray currents in the soil and metallic water pipes. If the stray current is large enough, step-voltages can cause problems for animals and human beings.

Conditions of imbalance are not constant and may start at certain hours of the day, lasting for a few hours. Time-stamped interval meter reading of all the energy users of each feeder can be used to determine how the loads are distributed amongst the phases at specific times the day. The coincident hourly demands can be used to determine how to redistribute the loads over the three phases of the feeder in order to reduce unbalanced loading.

Some of the necessary steps to accomplish this are the following. (Also, the assumption is made that load transfers across the feeder phases are permanent. No remotely controlled load transfer switches are used.)

- Obtain several days of synchronized time-stamped interval meter reading data for all the customers served by the feeder one is attempting to balance.
- Check whether the condition of imbalance repeats itself on a daily basis.
- Obtain the coincident load of each distribution transformer by summing the interval meter readings of all the customers served by the transformer.
- Obtain the coincident loads of each phase of the three-phase feeder.
- For each interval, determine the main contributors (distribution transformer loads) to the imbalance.
- Transfer some of the distribution transformer to another phase to achieve some improvement.
- For a period of 24 h check if this load transfer can sustain a certain degree of minimum imbalance.
- Several iterations may be required to accomplish this task.
- Once a degree of balance is achieved, continue to take synchronized time-stamped interval meter readings to verify the load balancing implementation that was based on historical interval metering data.

It is not possible to achieve completely balanced loading at all times. However this proposed method will have some merit by allowing one to come close to being able to proceed more accurately in the attempt to improve the balancing of loads.

5.6 Integrated volt-VAR control

5.6.1 Introduction

One of the useful applications for optimizing distribution feeder operation is *integrated volt-VAR* control. Capacitor banks and voltage regulators operate in a synergistic fashion to maintain a good voltage profile along the feeder and simultaneously control the line losses in an optimal fashion. The basic function of the capacitor is to relieve the source from having to supply the load reactive power. If the capacitor is placed at the correct strategic location on the feeder and switched in at the appropriate time, then the substation current flowing into the feeder is reduced because the capacitors provide the locally required inductive currents of the load. The voltage drop at the feeder and the conductor losses are reduced. Often, several capacitor banks strategically positioned and switched in and out at the appropriate times are required for the feeder.

Some of the difficulties are to locate the optimal placement for the capacitor banks, to determine the correct capacitor banks to switch and setting the correct switching schedules. Another problem is also how to correctly size the capacitor banks. One of the main problems is also the difficulty of obtaining accurate information on the load and voltage distribution along the feeder as a function of time and location. The switching in and switching out of capacitor banks are based on the estimation of anticipated line loading duration and voltage profiles from past studies, weather data, etc.

Synchronized interval meter reading of all the loads on a feeder coupled with voltage reading along the feeder can provide a correlation between loading profile and voltage profile along a feeder in real time. This type of information will help in the design of better switching schedules for capacitor banks and voltage regulator setting adjustments to optimize feeder wire losses and feeder voltage profiles.

Once a switching schedule is implemented, continuous synchronized meter reading and voltage data can be used to verify how well the switching schedule is designed.

5.6.2 VAR monitoring and control

If the currents in figure 5.15 can be expressed as follows

$$I_1 = I_{1\mathrm{re}} + jI_{1\mathrm{im}} \qquad I_2 = I_{2\mathrm{re}} + jI_{2\mathrm{im}}, \tag{5.11}$$

then the imaginary component due to the loads is equal to $j(I_{1\mathrm{im}} + I_{2\mathrm{im}})$ for scenario 1 operation. The imaginary component of the current is caused by the VAR component of the network and of the loads. For the period of time ΔT the substation bus and the capacitor bank have to supply approximately $-j(I_{1\mathrm{im}} + I_{2\mathrm{im}})$ to compensate for the VAR requirement. When the system is operated as depicted in scenario 2, the VAR requirement is less severe. The substation bus and capacitor banks have to be able to compensate jI_{1im} or jI_{2im}, whichever current is larger, but for a longer period of time $2\Delta T$. The voltage regulation on the feeder is also improved. In the following section a more detailed description will be provided of the methods to implement the optimization principles as described in this section.

The whole three-phase feeder is now replaced by a three-phase network with simple taps where each tap represents the single-phase equivalent impedance of the lateral subnetwork. This simplifies the complex feeder network dramatically and helps to determine through network modeling and analysis which actions to take to optimize the distribution of electric energy. Similar techniques can be applied to two-phase and three-phase laterals. The dynamic changes on the feeder are caused by the load changes. Synchronized interval data of kWh, kVAh and the rms voltage V_{rms} on the load side describe the feeder network dynamic changes with time. This type of information is useful for assessing the voltage distribution on the feeder and the losses incurred on the feeder and laterals. Control algorithms can be designed to improve the feeder load factor, load and voltage balance and also to construct a more comprehensive capacitor bank switching schedule. Daily, weekly and seasonal load profiles also help to adjust the controls to reflect the dynamic changes of the system.

5.6.3 The volt-VAR control application

For each time interval, the lateral shunt impedance can be calculated as a function of time valid for the defined interval ΔT. The shorter the time interval is chosen to be, the more accurate the results will be to describe the circuit dynamic changes with time. For the feeder in question all the equivalent load impedances at all the phases of the feeder can be calculated. The equivalent load impedances calculated by using equations (5.7)–(5.11) are based on the average value of the load within a specified time interval ΔT. If the circuit parameters of the three-phase feeder are known, the EMTP can be used to calculate the voltages and currents at all parts of the feeder circuit for a given set of input voltages at the substation bus and with the capacitor banks connected. For every new set of load intervals obtained from the smart meters, the currents and voltages on the line can be calculated. Hence the dynamic changes on the line are based on the data provided by the smart meters served by the feeder and not based on assumed load distribution models. For each state of the network as calculated using the EMTP software one can predict the expected outcomes of certain applications such as changing the capacitor banks' switching combinations, transferring laterals to a different phase, voltage regulator setting changes, etc. This provides the electric utility a means to obtain the most optimal operational gain at the feeders and the whole network served by the substation bus. The desirable gains are for instance improved load balance, a good voltage profile along the feeder, reduced circuit losses, etc. The EMTP results can be tested through actual implementation of selected control strategies on the network as predicted by the staged controls predicted by the EMTP calculations and the substation SCADA verifies the outcome of the application.

An operating AMI system which is synergistically coupled with a distribution SCADA can generate a wealth of information about the state of a distribution network. The ability to operate data gathering in synchronized controlled intervals and tag the circuit information to the collected data make the data more valuable. Advanced applications can be generated.

Specifically for volt-VAR applications, dynamic modeling of the network using smart metering data and tools such as EMTP allows the user to develop strategies to optimize volt-VAR controls. Verification of the results on a real system can be performed by operating the capacitor bank switching and voltage regulator control. The SCADA RTU and the AMI system provide the necessary feedback from the actual control action for verification.

The methods described in this chapter open the door to numerous other applications, such as voltage balancing, asset management, load factor improvements, etc.

IOP Publishing

New Technologies for Smart Grid Operation

Sioe T Mak

Chapter 6

The use of the utility communication network for monitoring power quality problems

6.1 Introduction

Recent advances in communication, digital electronics and computer technology have made remote monitoring of the electric utility distribution network possible. Numerous utility applications, such as automatic meter reading (AMR), demand response type projects and some distribution control functions, are already being implemented by many utilities and more added-value functions are being looked into or have already reached the pilot project stage. One of the most recent issues raised by the electric power industry is that of power quality monitoring. Power quality is one of the important issues that falls under the broad category of power delivery reliability.

The nature of power quality problems can be defined in terms of their detrimental impact on customers' appliances and devices and also on the supply network and energy monitoring devices. This chapter concentrates on the ability to locate power quality problems that are affecting customers and revenue metering as well as the ability to locate the culprits of harmonic pollution, which may or may not affect a broader segment of the network.

Without a good and reliable communication network, it is practically impossible to monitor each customer site for problems related to power quality because of the sheer number of sites to monitor in a timely fashion. As an example, the very stringent requirements imposed on the communication system for AMR opens the door for other applications. In particular, the recent trend of requiring synchronized time-stamped interval revenue meter reading makes the system more powerful for time correlation of events or conditions occurring in the power delivery network. Devices capable of monitoring power quality problems installed at strategic locations throughout the network and linked to the communication infrastructure will provide an invaluable amount of information that can be used to remedy situations

doi:10.1088/978-0-7503-1158-8ch6

that affect the quality of power delivery. Voltage swell and sag data from many sites, not merely evaluated statistically on an individual site basis but also correlated by time for all the sites being monitored, will be more valuable to the electric utility for setting up strategies to minimize their detrimental effects.

Other types of power quality problems are related to distortions of the 60 Hz voltage and current waveform. Numerous papers have been written on the effects of harmonics on losses in electric power networks and major appliances. Many sensitive digital electronic devices are also affected by distorted voltages. Digital clocks, relying on the 60 Hz voltage zero-crossings for their proper operation, are also affected by recurring transient spikes on the voltage waveform. Other malfunctions are caused by memory contamination of these devices due to burst transient phenomena.

The degree of severity of the voltage distortions can be such that many customers' devices are affected simultaneously due to a main culprit. Monitoring the problems at the point of common coupling quite often produces results which point to the source of the problem. A more commonly occurring power quality problem, caused by locally generated harmonics and only affecting devices in its immediate vicinity, is more difficult to locate.

Revenue meters are also prone to being affected by locally generated distorted voltages and currents. It is difficult to quantify the cumulative loss of revenue due to harmonic pollution. This chapter is an attempt to explain and discuss the main issues related to the detrimental effects of severely distorted voltages and currents on revenue meters and how the existing communication system already used for remote metering can be used to quickly determine the sites affected by harmonic pollution.

In addition, there is the question of which metric can be used to quantify distortion and what data should a power quality monitoring device generate, time-stamp and make available for data gathering using the communication network.

6.2 The root causes of power quality problems

When power quality problems arise at metering sites, meter reading accuracy may be compromised. Loads can cause fluctuating voltages at the point of connection to the network which will be different from the source voltage. This voltage can also be distorted by some types of loads at the point of connection. The energy consumption is generally also monitored at the same point of connection.

One of the categories pertaining to voltages in the area of power quality issues is voltage stability. Large steady state loads that cause network overloading and are subject to frequent switching in and out during short intervals, the loss of voltage due to re-closer operations, etc, cause voltage dips or swells that may have undesirable effects for the electric users. To determine the causes of these voltage problems, one assumes that the voltage at the point of use maintains its sinusoidal wave shape and frequency except during a short period after a switching action. The current standards and recommendations already address these situations.

Switching generates transient voltages that may cause problems for equipment and devices. Recurrent transients, burst transients caused by power electronic switching devices, arc furnaces, arc welding, etc, occur quite often in the electric network and cause voltage and current distortions. Numerous studies on the hazards and detrimental effects of harmonics on the electric network, equipment, etc, have been published. Other sources of power quality problems are high impedance faults, surface discharges on polluted insulators, bad contacts, etc.

In addition to affecting appliances and metering devices, there is also the possibility that the quality of power can cause communication problems. Power line communication (PLC) technologies are especially vulnerable. Communication signals are contaminated by burst harmonics. Recurring transients sometimes cause damage to the memory or the electronic hardware of the communication devices. A great deal of effort and money have been spent to make communication systems reliable and less sensitive to power quality problems. Severe communication impairments can be used as alerts that power quality problems are occurring on the communication channel being used. New generation electric meters—digital electronic based meters—are purported to have the ability to measure kVAh, kWh, power factor, demand, etc. This leads to the possibility to more comprehensive and novel billing strategies. The existing standards require that these meters guarantee a certain degree of accuracy when calibrated with clean sinusoidal voltage and current waveforms. Intuitively it is clear that a 'clean' ac electric power source comprises a sinusoidal voltage that maintains a constant amplitude and frequency under various loading conditions. However, this assertion deviates from reality since the source and the power delivery network have impedances which may produce phase differences and decreased voltage amplitudes at different locations on the network.

Studies are available on the effects of distorted voltage and currents on electric revenue meters and kVAh metering. However, the lack of a good definition of what distortion means can become a problem. An industry accepted form of standardized distortion recognized by the various testing laboratories has yet to be defined. In this chapter, an attempt is made to clarify and describe the main issues and to recommend certain steps needed to consolidate future efforts in order to arrive at testing standards that are logical and that are possible to implement.

6.3 Descriptions of some types of waveform distortions

Transients. A transient is a subcycle overvoltage wave in an electric circuit generating a disturbance of the power voltage waveform (figure 6.1(*a*)).

Impulsive. The duration can range between less than 50 nanoseconds and over 1 millisecond and can reach magnitudes of several per unit depending on the disturbing source and on the voltage withstanding characteristics of the system.

Oscillatory. The duration is between 0.3 and 50 milliseconds in a 60 Hz system. Its frequency ranges between less than 5 kHz and 5 MHz and the magnitude can reach up to 8.0 unit^{-1}.

Short duration variations (sag and swell). Sag is a momentary under-voltage at fundamental frequency lasting from half a cycle to one minute, as seen in figure 6.1(*b*). Swell is a momentary overvoltage at fundamental frequency of similar duration to a sag. See figure 6.1(*c*).

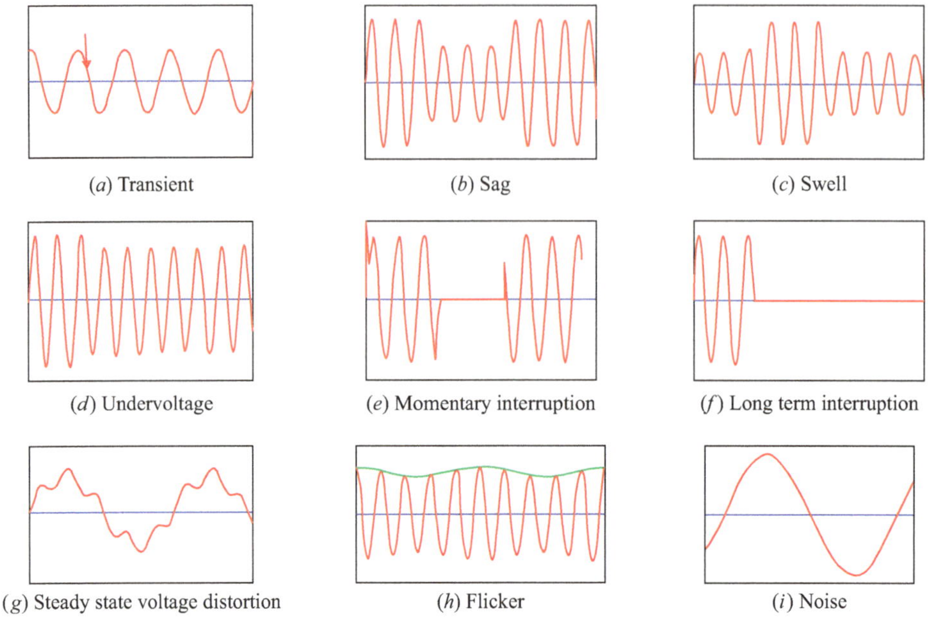

Figure 6.1. Some types of common waveform distortions.

Instantaneous swell or sag. The duration is between 0.5 and 30 cycles with magnitudes ranging between 0.1 and 1.8 unit^{-1}.

Momentary swell or sag. The duration is between 30 cycles and 3.0 s with magnitudes ranging between 0.1 and 1.4 unit^{-1}.

Temporary swell or sag. The duration is between 3.0 s and 1.0 min with magnitudes ranging between 0.1 and 1.2 unit^{-1}.

Long duration variations:

Overvoltage. The duration is longer than 1.0 min with magnitudes ranging between 1.0 and 1.2 min.

Undervoltage. The duration is longer than 1.0 min with magnitudes ranging between 0.80 and 0.9 unit^{-1}, as illustrated in figure 6.1(*d*).

Interruptions. An interruption is the complete loss of voltage for a period of time.

Instantaneous. The duration is between 0.5 and 30 cycles.

Momentary. The duration is between 30 cycles and 3.0 s. See figure 6.1(*e*).

Temporary. The duration is between 3.0 s and 1.0 min.

Long term. The duration is longer than 1.0 min. See figure 6.1(*f*).

Distortion. A waveform distortion is any deviation from the nominal sine wave of the ac line voltage or current. The spectrum ranges between 0.0 Hz and the 100th harmonic, the harmonic strength ranges between 0.0% and 20.0% for the voltage and for the current the strength ranges between 0.0% and 100.0%. An example of a severe voltage distortion produced by a non-linear load with high 5th harmonic current is shown in figure 6.1(*g*).

Flicker (voltage fluctuations). Flicker is a variation of the input voltage as illustrated in figure 6.1(*h*), sufficient in duration to allow visual observation of a change in electric light source intensity. It is intermittent and its magnitude ranges between 0.1% and 7.0 %.

Noise. Electric noise is unwanted electric signals, see figure 6.1(*i*), which may produce undesirable effects in the circuits of the control systems in which they occur. They are random and can be very short and intermittent or continuous, with magnitudes reaching up to 1% unit^{-1} of the fundamental voltage.

These definitions are fine, but the question remains: what does one do with them? A fundamental need is to understand the sources of the distortions, to characterize them and to decide how to use the information. In order to identify these distortions, a method for processing the distorted waveform using a remote device is needed. The result of the processing is the piece of information that will be communicated to the central computer. Would total harmonic distortion (THD) meet the criteria?

There are cases where a large non-linear load is served by a sinusoidal voltage source. The distorted current creates non-sinusoidal voltage drops. Adjacent neighboring electric customers suffer from the distorted voltage. Sometimes the non-linear load is small enough not to cause spillover to adjacent customers and yet is severe enough to cause local problems. This case is the most difficult to detect.

When the load varies with time, the distortion is not constant with time. What one measures today may not be a reflection of tomorrow's measurements.

6.4 Problematic considerations

Many of the published results of investigations on the effects of voltage distortions on electric metering devices have been primarily concerned with the difficulties in defining and measuring the energy entity 'kVA' under harmonic distortion conditions. Distinctions have to be made between distorted voltages serving linear loads and voltage distortion caused by non-linear loads. Issues of unbalanced loading on three-phase systems are also problematic.

All of the distortions used to check metering calibrations are due to steady state harmonics. Measurement results indicate that under conditions of distortion due to steady state harmonics, kWh measurements do not cause appreciable loss of metering accuracy. The problem is with kVAh measurements because it is not clear how one should define kVA under distorted conditions of the voltage and current waveforms.

One major aspect that has been absent in all previous investigations is the case when the distortion contains non-integer harmonics. The complexity increases dramatically when the relationship between the source voltage and the load current is non-linear. A non-steady state dc component in the load current adds to the complexity of the problem.

Several important questions can be raised in this case:

1. What are the effects on electromechanical meters that use a Ferraris disk?
2. What are the effects on electronic meters?
3. Which components or parts of the meter are involved in metering inaccuracies?

4. Do concepts such as kVA, power factor and reactive power still make sense if these items are used for electric energy metering and billing?
5. Does filtering affect actual energy consumption?
6. Can the distortion be categorized in several classes that can be used for calibration purposes?
7. What types of remedies can be applied to the sources of distortion?

6.4.1 Causes of inaccuracies

To appreciate the problem that we mentioned above, let us take a fairly simple example for purposes of illustration. A simple induction disk electric meter operates on the following principles. An ac passing through the current coil of the meter generates an alternating magnetic field that crosses the rotating disk perpendicularly, see figure 6.2(*a*). The induced electromotive force (EMF) in the metallic disk causes

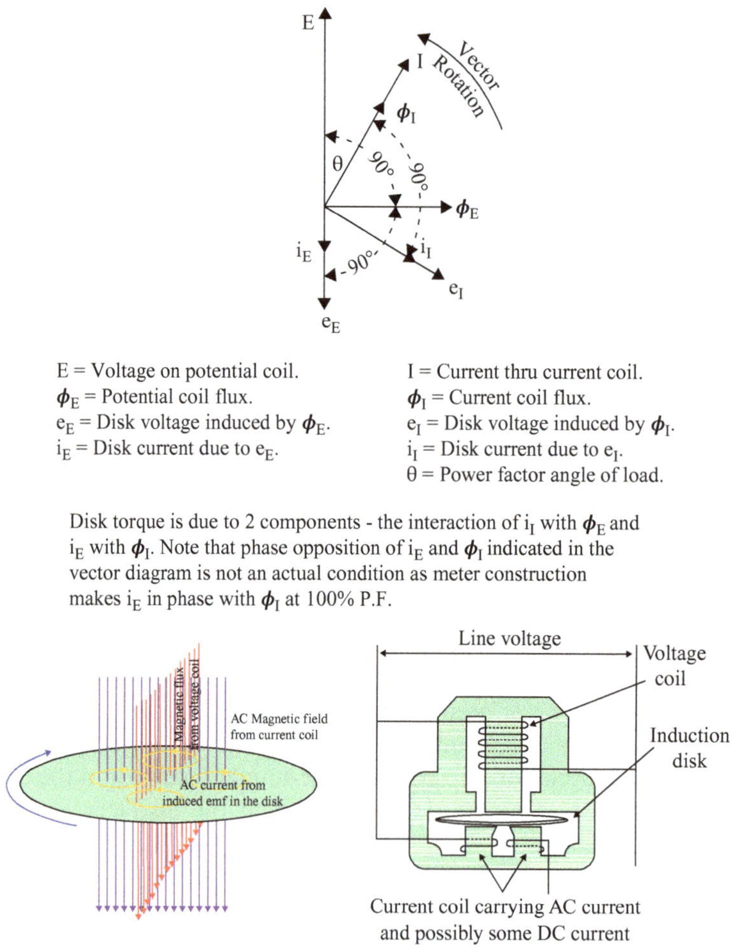

Figure 6.2. Diagram showing the direction of rotation of the induction disk.

an ac to flow in the disk, which reacts with the magnetic flux generated by the voltage coil, figure 6.2. The sum of all the elemental Lorentz forces generates a net mechanical torque which rotates the disk.

(a) *Retardation torques.* If in addition to the ac a dc is also present in the current coil, a dc flux is also generated and superimposed on the ac magnetic flux. The disk rotation generates an EMF due to motion in the disk. This EMF causes current flow in the disk and retardation torques.

(b) *Harmonic torques.* Additional torques can also be produced by steady state harmonic currents from non-linear loads on the customer side and these can be in the same or in the opposite direction to the fundamental frequency torque, depending on whether the harmonic currents are positive- or negative-sequence. Slowing down the disk rotation essentially means a loss of revenue. Likewise, continuing positive-sequence currents can impose accelerating torques on the disk that may affect customers' metering. These findings have been identified by several investigators and reported in the literature. Moreover, studies conducted by Louisiana State University indicated that a loss of accuracy of as much as 8% is possible when both the voltage and currents contain harmonics.

(c) *Transducers.* Electronic meters are coupled to the power line through transducers. The most common types of coupling devices are potential and current transformers. These transformers have a certain degree of accuracy guaranteed when subjected to 60/50 Hz voltages and currents. When subjected to distorted voltages or currents, it is not clear whether the voltage and current ratios and inherent transformer phase shifts are still within the certified range of accuracy for pure sinusoidal conditions. A dc component in the voltage or current may cause core saturation. High frequency harmonics are transferred through capacitive coupling. The skin effect increases the winding impedances. High voltage spikes may affect the contents of memory devices or the display used in the electronic meter.

6.4.2 Examples of extreme waveform distortion

Some examples of extreme distortion on voltage and current waveforms are shown in figures 6.3–6.6. The voltage of figure 6.3 was taken at the service box of a three-phase supply serving a pump motor with a variable speed drive. The line-to-line voltage was 480 V and measurements at the medium voltage level did not indicate any cause for concern of damaging spillover to other customers.

The voltage depicted in figure 6.4 was captured at the installation of another variable speed drive. The distortion is completely different to that depicted in figure 6.3.

Figure 6.5 depicts the unfiltered current to a six-phase power rectifier serving a 100 kW TV transmitter load. The smoothing filter capacitors were never installed for some unknown reason.

In figure 6.6 a load current rich in the third harmonic from a small industrial plant is shown.

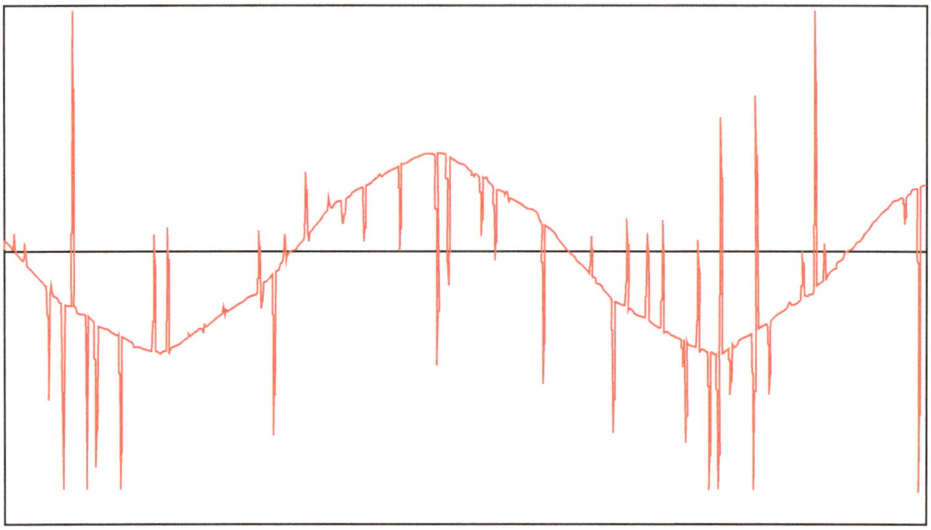

Figure 6.3. A severely distorted voltage waveform at a pump motor installation.

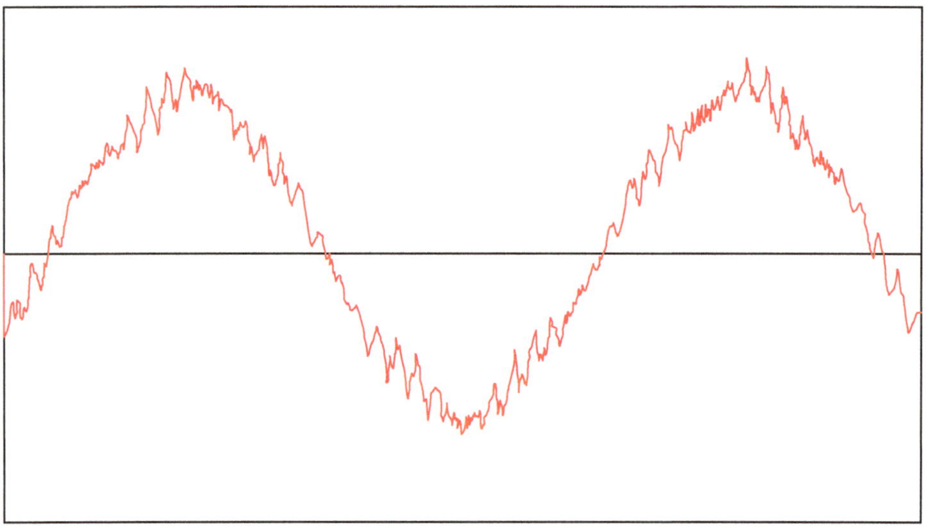

Figure 6.4. Another example of voltage distortion at a frequency drive installation.

Figures 6.3 and 6.4 do not exhibit steady state conditions and are difficult to characterize in terms of Fourier series and in essence represent a distorted periodic waveform as the superposition of a series of higher-order frequency components (harmonics) and a fundamental sinusoidal component. Figures 6.5 and 6.6 are more amenable to this kind of analytical representation.

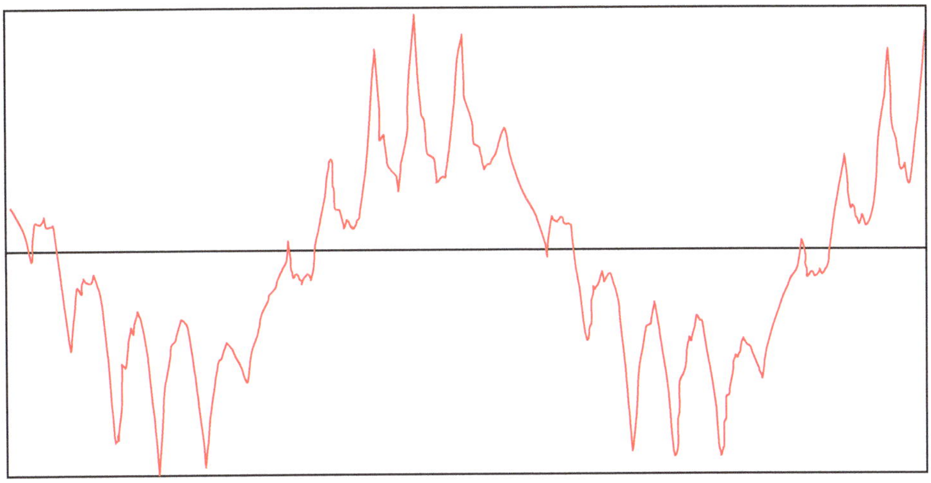

Figure 6.5. An unfiltered current to a six-phase power rectifier.

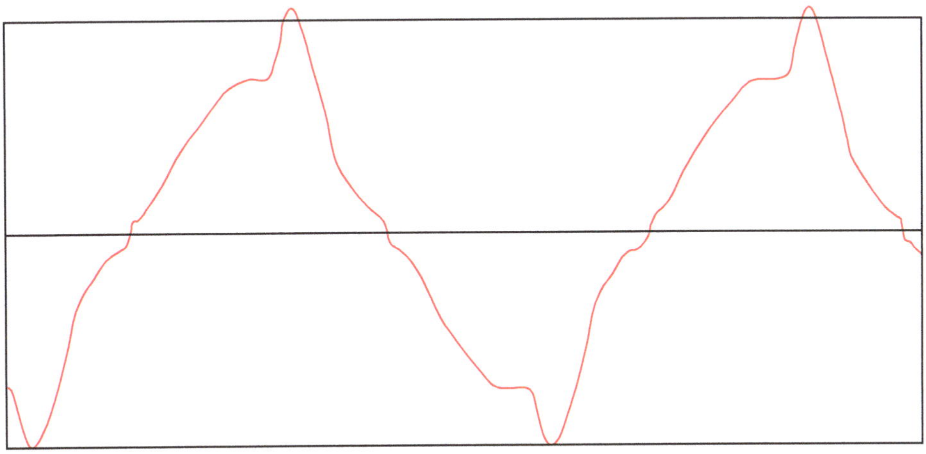

Figure 6.6. Significant third harmonic distortion in a current waveform.

6.5 Standards issues

The existing standard on power quality in the USA related to metering accuracy has been discussed by many professionals. Their concerns are primarily based on distortions due to integer harmonics of the 60 Hz waveform. However numerous field investigations have indicated that many distorted waveforms are rich in non-integer subharmonics and behave like steady state waves for certain intervals of time. A few such waveforms were shown in figures 6.3–6.6.

There are suggestions for introducing new rate structures based on kVA rather than on kW. These suggest penalizing the customer for a bad power factor. For a steady state sinusoidal voltage and current, kVA and kW are coupled by the concept of the power factor. The problem becomes extremely complex when both voltage and current are distorted. Under non-sinusoidal conditions and three-phase unbalanced conditions, engineers and scientists involved in the study of power entities have yet to agree what kVA means and whether there is even a correlation between real power and kVA.

At this moment it is not clear how much revenue is lost due to inaccurate meter registration due to severe waveform distortion within an electric utility. Locations or institutions performing calibration of electric energy meters under distorted voltage and current conditions are scarce and are not generally available or known. It is also not clear what *standard distortion* means and what a good measure of accuracy is.

The time has probably arrived to develop some generally accepted standards of distorted waveforms. For example, in the area of transient voltage testing, standard test waveforms such as the 1.2×50 μs voltage impulse wave and the 10×20 μs current impulse wave for testing the dielectric integrity of electric devices were adopted a long time ago. Yet these test waveforms are not necessarily replicas of the transient waveforms measured in the electric networks. However it has been proven repeatedly that withstand levels to transients as defined by these tests using the indicated waveforms provide a guaranteed level of survival against normally occurring damaging transients.

The same effort could be applied to arrive at some standard distortion of both voltage and current waveforms. Voltage and current waveforms are not independent of each other and using modern day data acquisition systems gathering these types of data is not difficult. References [10–12] described some of the techniques used to replicate distorted waveforms in a testing laboratory for meter calibrations.

6.6 The role of communications

With a communication infrastructure already in place, the question arises of what is the nature of the information needed to bring back to the dispatch center computer.

One possible method, pending the development of new meaningful metrics to measure power quality, is the use of a contiguous sequence of short time monitoring intervals ΔT. For each interval of ΔT the THD can be computed. For a contiguous number N of such intervals the maximum THD_{max}, the minimum THD_{min} and the average THD_{avg} can be computed. This method allows one to identify the degree of bursts occurring per unit time. If THD_{avg} is very close in magnitude to THD_{max} and THD_{min} then this simply means that there are no minima or maxima. If THD_{avg} is close to the magnitude of THD_{max} then this simply means that there are more maxima then minima.

The interval $(N \times \Delta T)$ is time-stamped and the THD data are retrieved from the remote monitoring site using the communication system. ΔT can be as short as one or a few cycles of the 60 Hz wave and $(N \times \Delta T)$ can be as long as 15 or 30 min. To reduce the amount of communication bandwidth used, a number of consecutive intervals $(N \times \Delta T)$ can be stored in the memory of the monitoring device and retrieval of the stored data is performed on a daily schedule.

To take advantage of the communication network it is also possible to simplify the time correlation of the collected data by service territory. The following method can be used. All power quality monitoring devices connected to a distribution feeder or phase of a feeder can be assigned to a functional group address which can be communicated to by a single group command. The functional group data do not only provide time correlation, but also correlation by location served by the same supply or source.

The same technique can be applied to the monitored voltage. By time correlating the time-stamped collected data for a certain service territory, feeder or network section served by a distribution bus, one can decide whether the problem is local or widespread. The magnitudes of the monitored data also provide an indication of where the culprit of the harmonic pollution is located. Power line types of communication technologies are affected by severe power quality problems. Unusual degradation of the communication performance on a physical channel will serve as an alert of power quality degradation. Strategically placed advanced electronic meters capable of monitoring THD and voltages can be used to collect useful information. Also PLC impairments could be used as an alarm to pinpoint distribution network areas where power quality problems occur.

6.7 Conclusions

1. The effects of waveform distortion of the 60 Hz voltage and the current on metering devices and the lack of test standards have been discussed.
2. Loss of revenue due to inaccurate energy metering will offset the benefits gained by having an AMR system. Conversely, metering inaccuracies due to waveform distortion effects as described in this chapter can sometimes also affect customers' devices.
3. A communication system for AMR, hardened to harmonic distortion, can be used to pinpoint localized power quality problems.
4. Some sophisticated electronic meters already exist, which have the capability to monitor THD. For the time being, they may be candidates for monitoring power quality problems by strategically scattering them throughout the distribution network.
5. For PLC systems, a sudden drastic drop in communication performance on certain physical communication channels can be used as an alert that a power quality problem exists on that part of the electric network.
6. There is definitely a need for a standardized distorted waveform that can be used for revenue meter calibration purposes and which can be replicated in testing laboratories. In addition, a metric has to be defined as a measure of metering accuracy when operating under distorted voltages and currents.
7. Consequently, as we improve our understanding of the effects of voltage distortion on meter accuracy, and in the search for answers to the above questions, we realize the great potential for using the communication system for AMR in conjunction with data acquisition equipment strategically positioned on the distribution network to better identify power quality problems.

IOP Publishing

Sioe T Mak

Chapter 7

Different types of communication systems for utility applications

7.1 Introduction

To treat the subject of the different types of communications systems in detail is beyond the scope of this book. However, there are unique physical aspects to the various technologies which form the basis of communication systems. The physical communication medium, the carrier frequency used and the propagation characteristics of the communication medium show important differences that will decide what the fundamental limitations of a system will be for certain utility applications. These issues will be explored and explained in great detail.

The utility communication network can be categorized into several generic types:

- power line based communication (PLC) technology;
- radio frequency (RF) communication technology;
- hybrid systems, which are a mixture of hard-wired and RF networks;
- communication technologies that use other types of hardwired networks, such as the public telephone network, a dedicated fiber optic network, etc;
- broadband systems and satellite based communication.

Distinctions can also be made between 100% two-way communication systems and 100% one-way systems. Hybrids of one-way systems coupled to a two-way trunk system also exist. In this chapter detailed discussions will be devoted to the strengths and weaknesses of each type of technology in the light of possible future utility type applications of added-value functions. Primary consideration will be given to electric utility applications and, where appropriate, the service extensions provided to other utilities such as gas and water utilities. Some insight about the key issues have already been provided in chapter 6.

doi:10.1088/978-0-7503-1158-8ch7

7.2 Power-line based communication technologies

PLC at the transmission voltage level has existed for a long time and is primarily used for protective relaying purposes. In this book the discussion primarily focuses on PLC technologies which use the medium voltage distribution network as their propagation medium.

PLC technology uses the power delivery network as its communication medium. It can be a one-way or two-way system and in most cases it operates in the utility medium voltage distribution and low voltage networks. Some PLC technologies only operate at the service voltage level.

One should remember that the primary function of the distribution network is to deliver electric energy to the consumers at a constant frequency and nominal voltage level. A basic understanding of the physical layout and physical properties of the distribution network is necessary to appreciate the propagation characteristics and behavior of some of the communication technologies in that environment.

A typical electric distribution network consists of a distribution substation transformer, which steps the voltage down from a transmission or subtransmission level to a three-phase medium voltage level. This medium voltage design level ranges between 4.0 kV and 34.5 kV. There may be several three-phase feeders emanating from the medium voltage distribution substation bus. They can be over-head wires or underground cables serving different types of customers in the service territory and these lines can be three-phase with or without a neutral. Numerous three-phase or single-phase taps and laterals are provided. The main feeder lengths range between 2 and 15 miles and in some rural areas 60 mile-long feeders are not uncommon. These feeders may have segments at different voltages and be coupled to each other through delta-wye, wye-wye or autotransformers. Energy delivery to the customers is provided by single-phase step-down transformers at 240 V/120 V three-wires, or three-phase transformers at 208 V/120 V four-wires or 480 V/277 V four-wires. At selected locations on the feeders or laterals, protective devices and switches are installed. Line voltage regulators and capacitor banks are used to maintain a good voltage profile along the feeder circuit and to minimize losses in the network.

One should stress the fact that a well-designed distribution network is intended to deliver electric energy at a frequency of 50 or 60 Hz while maintaining a fairly constant voltage to the consumers in the most reliable fashion. It was not intended as a communication medium. However, recent technological developments indicate that the possibility exists of using the power delivery infrastructure as a communication infrastructure. This eliminates the need to invest in a communication infrastructure and, in addition, the electric network reaches practically 100% of the population compared to other types of utility services.

Other issues of general concern are where the outbound signal is injected and whether customers tapped between line-to-line and line-to-neutral off the feeders can receive the outbound signals. If a distribution substation transformer winding is connected in grounded wye on the distribution voltage side, then injecting the signal in the substation transformer neutral will not be detectable by transponders across distribution transformers tapped line-to-line off the distribution feeders.

Several different types of one-way and two-way communication technologies exist which operate at the medium voltage and service voltage levels. Other technologies operate at the service voltage level only. To proceed with this discussion in a more systematic fashion, the PLC technologies mentioned above are categorized as follows:

- one-way outbound only at the medium and service voltage levels;
- one-way inbound only at the medium and service voltage levels;
- two-way medium frequency at the medium and service voltage levels;
- two-way low frequency at the medium and service voltage levels;
- broadband, high and medium frequency carriers at the medium and service voltage levels.

7.2.1 One-way outbound only

The power line *one-way outbound only* (figure 7.1) communication system has been used for quite some time for load control. Here, the discussion of this technology is primarily intended to show the reader why the signal propagation is so reliable. The primary node is the outbound transmitter and it is coupled to the subtransmission line or to the individual distribution substation bus. The transmitter injects low frequency signals (200–400 Hz) to modulate the low frequency 60 Hz or 50 Hz power voltage wave, which can be encoded to carry information. These signals

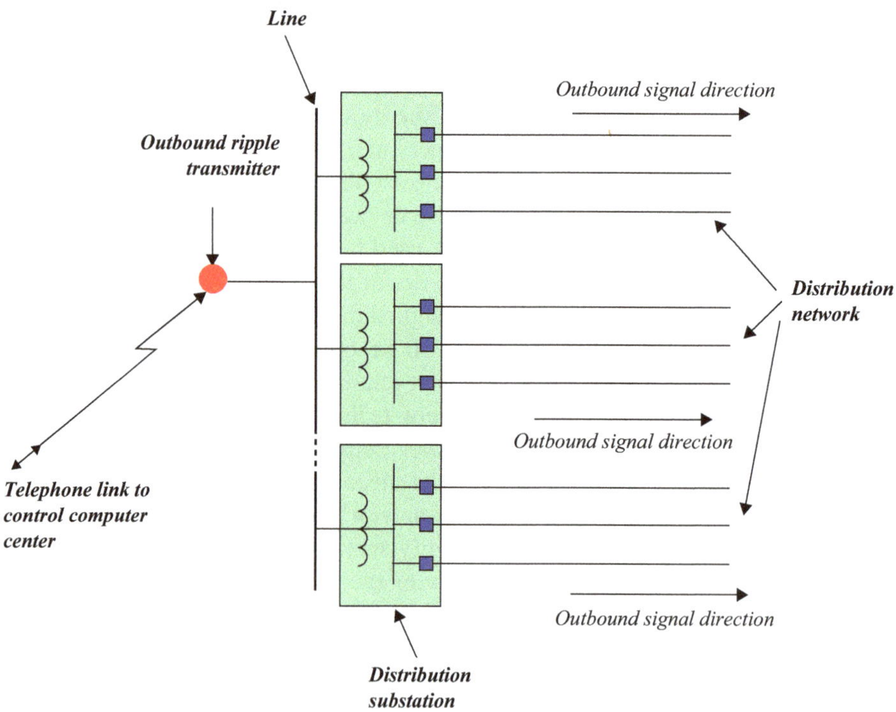

Figure 7.1. One-way outbound transmitter connection diagram.

propagate through the electric network across all types of medium voltage step-down transformers and across distribution transformers to the low voltage network in a very reliable manner. Capacitor banks and underground cables do not, in practice, attenuate the signals and normally encountered feeder circuit lengths do not cause standing wave problems. The reliability of this communication technology can be attributed to the low frequency signals used. These signals are well within the range of the natural frequency response of the medium frequency distribution network caused by a short transient perturbation.

The well-known one-way *ripple system* has been used in some parts of the USA and is used widely in other countries for load control. In some European countries this technology is used to download tariff changes to electric meters, to download pricing signals, to control streetlights and to control water heaters and space heaters. A commercially available two-way version of the ripple technology has yet to be developed.

Expanding this technology for two-way applications is possible while still retaining its original one-way application for load control. The inbound communication has to be developed and designed. As for the outbound signaling technology, the existing technology is solid and reliable and can still be used.

7.2.2 One-way inbound only

The power line *one-way inbound* (figure 7.2) communication technology does not have any outbound transmitters to communicate with remote devices. Since there is no way

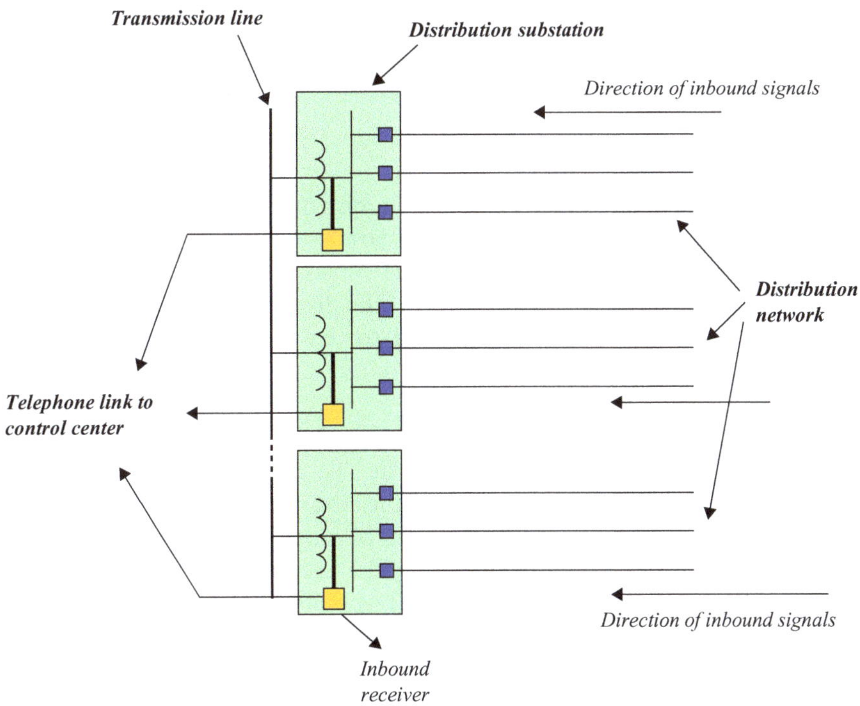

Figure 7.2. One-way inbound communication receiver connection diagram.

to control and schedule the inbound transmission, the transponder generates the inbound signals in the low voltage network in an unscheduled and unsolicited fashion. The signals propagate across the distribution transformers to the medium voltage circuit and the inbound receiver captures the signal from the current transformers at the distribution substation. The frequency used for communication is very low.

This technology is primarily used for revenue metering and to some extent it can be used to detect outages at remote sites. The inbound transmitters are located at the electric meters at the service voltage level. Metering data from all metering transponders are transmitted in an unsolicited inbound fashion over the low voltage network and the feeder network to the inbound receivers, which are located at the distribution substations. The transmission of the inbound messages from the different transponders cannot be synchronized with time and are allowed to collide. Multiply retries with random time delays are required. At least a few will slip through each time. They are received and decoded by the inbound receiver and subsequently transmitted to the central control computer. Due to the multiple collisions, not too many metering data can be received per unit of time. The primary node is the inbound receiver and is located at the distribution substation. All metering data are eventually collected after a certain time period and sent to the central control computer. This system is still in use by many electric cooperatives for electric billing purposes.

Conversion into a two-way system by adding an outbound transmitter will require a complete change in the transponder and communication system design, and the transponders already installed for one-way inbound only and operating may have to be completely replaced.

7.2.3 Two-way systems

Medium frequency carrier

Two-way medium frequency power line carrier systems typically operate only in the distribution network. The medium frequency carriers operate at frequencies ranging between 5.0 and 15.0 kHz. The primary node consists of an outbound transmitter and an inbound receiver, which are located at the distribution substations as shown in figure 7.3. The transponders are coupled to the electric meters in the service voltage network. The load control or other devices are also located in the service voltage network.

Typical applications are remote meter reading, load surveys, demand metering, load management, distribution automation, etc. To assess this technology, several basic issues have to be considered. The use of the electric network for communication at a carrier frequency, which is several orders of magnitude larger than the power frequency, requires special considerations. The electric distribution network is designed to transport electric power at 50 Hz or 60 Hz. From a communication frequency standpoint, a transmission line is considered long if its physical length is close to a quarter wavelength of the carrier frequency. To obtain an idea of what a long line is at certain frequencies, the quarter wavelengths for various sinusoidal carrier frequencies are shown in table 7.1.

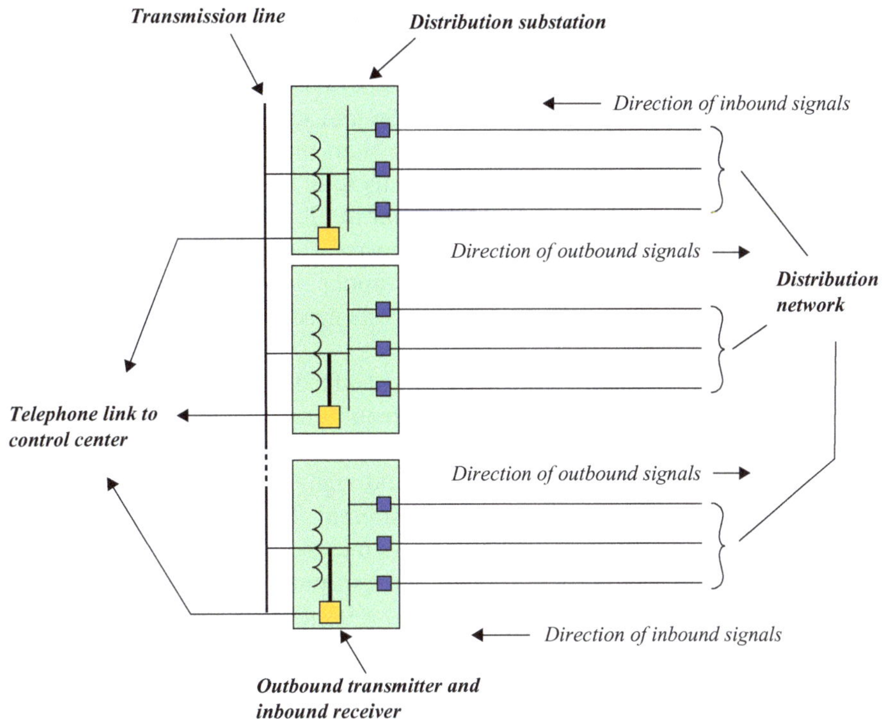

Figure 7.3. Outbound transmitter and inbound receiver connection diagram at the distribution substation.

Table 7.1. Ranges of quarter wavelengths, in miles an aerial lines at various carrier frequencies in Hz.

Carrier frequency in Hz	Quarter wavelength in miles
60	776.74
1000	46.60
5000	9.32
10 000	4.67
50 000	0.932

For a long line at a certain carrier frequency, one cannot use a lump parameter circuit for analysis and study. Distributed parameter network models are the appropriate models to use. Multiple reflections at mismatched junctions can generate standing wave patterns on the line. The nodes of the standing wave pattern on the line are points of the carrier wave that are at the lowest energy level. At the medium frequencies listed in table 7.1, the quarter wavelength for aerial lines ranges between 1 and 50 miles. The lengths of commonly encountered distribution feeders are well within the range of the quarter wavelength of these carrier frequencies. Hence, one can expect standing wave problems to exist due to multiple reflections. Transponders located near or at a nodal point of the standing wave will see very low signal strengths. Increasing the signal power is not going to help unless the carrier frequency is changed. To eliminate this problem repeaters are installed near the standing wave nodes on the line to amplify the signals.

Capacitor banks for power factor correction and underground cables form low impedance shunts at frequencies much higher than the power frequency and they will cause signal attenuation. Also, on sufficiently long lines, signals that are injected on a particular phase of the network will appear at all three phases at the other end of the line due to inter-capacitance coupling between the phase wires or intermediate step-down transformer windings. For some applications, this phenomenon can become a problem. It is especially important if one needs to know the phase location of the transponder in the electric circuit.

Another problem is 'signal spill-over' from one feeder to another feeder through the distribution substation transformer or from one distribution substation to another nearby distribution substation served by the same transmission line. The feeder-to-feeder signal spillover causes inbound signals from two different transponders signaling almost simultaneously at different feeders to interfere with each other, and hence mutually destroy each other's message.

The inbound signals are power limited and are most likely heavily attenuated. Proper isolation of the capacitor banks for those frequencies, line conditioning and installing repeaters, in particular for the inbound communication, can overcome some of the signal attenuation problems for specific network configurations. If the distribution networks undergo substantial circuit expansion in the future, or new capacitor banks are added, then filtering and repeater signal amplification and location changes are also required. To implement these changes is not always easy and can be expensive.

Low frequency carrier

This technology basically uses the power frequency voltage for the outbound carrier and the load current for the inbound carrier. The signal can be injected and superimposed on the power frequency voltage or current. Another method is to cause a slight distortion at specified portions of the waveform of the power frequency voltage or load current. The frequency of the injected signal is usually between 200 Hz and 500 Hz. The technology, which uses controlled distortion for outbound communication, accomplishes voltage modulation by drawing a controlled current pulse near the bus voltage zero crossing. This incremental pulse type load causes a voltage perturbation at the distribution bus and is transient oscillatory. This voltage perturbation is the distribution network transient oscillatory response and the transient oscillatory frequency ranges between 200 Hz and 500 Hz in 60 Hz and 50 Hz systems. The low frequency inbound signal can be injected by the transponder in the service voltage circuit. Another method is for the transponder to draw a controlled current pulse at the service voltage level. Extraction of the inbound signal is from the current transformers at the distribution substations.

These transient oscillations propagate throughout the whole network and decay within less than half a cycle of the power frequency voltage. This is one of the main reasons why the low frequency signal propagation does not suffer from standing wave problems, attenuation by capacitor banks and underground cables. Field studies show that distances of up to 60 miles pose no problem for communication. These long distribution feeders are commonly encountered in rural areas.

Another important feature of this technology is the phasor specific characteristics of this technology. For a three-phase system, one can identify three basic phase voltage phasors, $\bar{V}_{AN}$, $\bar{V}_{BN}$ and $\bar{V}_{CN}$ for the line-to-neutral voltages. The line-to-line voltage phasors $\bar{V}_{AB}$, $\bar{V}_{BC}$ and $\bar{V}_{CA}$ are linear combinations of the three base phase-to-neutral voltage phasors. When an outbound signal is injected at the distribution substation bus on a specific phase voltage, that signal 'follows' the phasor everywhere in the system across different three-phase step-down transformers of all possible different winding configurations or single-phase transformers without seeing appreciable signal attenuation. The inbound signal generated at the transponder follows the load current all the way to the distribution bus.

This unique feature allows one to use the *communication path* concept. We use the following example as an illustration.

A search algorithm is used to 'find' a transponder. For example, the transponder has a serial number NKNKNK. By sending a query to this transponder on all possible phases to respond back, it is found that the transponder responds to the query when the outbound message is send on phase V_{AN}. The inbound message is extracted from the current transformer of phase A.

- The communication path parameter is now defined as follows:
  i. To reach transponder NKNKNK, the outbound path is phase V_{AN}.
  ii. The inbound message can be extracted from the bus current transformer of phase A.
- For all subsequent transactions to transponder number NKNKNK, one uses the defined path.
- In the database, each transponder number has a unique path parameter found by the search algorithm.

The use of low frequency signals implies that the baud rate is low. The future requirements for gathering large amounts of information from remote sites can only be handled by using simultaneous multi-channel communications, reducing all unnecessary information in the message structure and a high degree of communication performance. One well-known technology can use up to six simultaneous communication channels per physical path per feeder per distribution substation bus. This means that six remote devices can respond simultaneously as a response to a single group outbound command per feeder per distribution substation bus.

Medium frequency power line carrier operation in the service voltage network
Some technologies are available which only operate at the service voltage level below each distribution transformer. Possible applications are for home automation, alarm systems, meter reading, etc. Examples are the X-10 systems, the Intellon CEBUS, the Echelon system, etc. Typical operating frequencies are in the range of 90 000 Hz and higher. In some metering applications, a data concentrator is coupled

to the low voltage network bus of the distribution transformer and serves as a primary node. The data concentrator sends all information through a telephone link to the server and subsequently to the customer service computer center.

The CEBUS system design is based on peer-to-peer communication. Each node in the network is an intelligent terminal and a peer, and communication to other nodes can be initiated at any time and repeated as many times as necessary and only stops transmitting when an acknowledgment from another node is received. Collision detection and collision avoidance capabilities are required for each node.

In many countries (other than the USA), step-down transformers have very large power ratings and range between 0.5 MVA and 2.00 MVA. The low voltage network is enormous and serves a large number of customers. There are only a few step-down transformers per medium voltage feeder. There are not too many distribution step-down transformers per service territory served by a distribution substation transformer at the medium voltage level. Hence not too many data concentrators linked directly by telephone to the central computer per distribution substation are required.

In the USA the distribution transformers typically have much smaller power ratings. A medium voltage feeder typically serves 20–40 distribution transformers. Hence low voltage network power line carrier systems are less popular for meter reading in the USA because of the large number of data concentrators (main nodes) and special links to the server that are required.

Other field studies also indicate that distribution transformers do not prevent signal spill-over from one low voltage network through the distribution transformer over the medium voltage circuit and through another distribution transformer to another low voltage network. If both low voltage networks have communication activities, some slow-down due to mutual interference can occur. Capacitor banks and underground cables attenuate the signal significantly and for practical purposes cannot be used to control distribution network devices. However, they are excellent candidates for home automation and unique local services.

7.3 Two-way telephone and hybrid telephone short hop radio systems

Two-way telephone systems for reading gas and water meters have been in existence for quite some time. Some systems for reading electric meters have not taken off into large systems. The telephone network is really designed to handle a large number of personal users, allowing random access and unrestricted utilization of the communication network. For large scale scheduled operations where global commands have to be broadcast to a large number of transponders at remote sites, the telephone system may cause unacceptable time delays to private users who need to obtain access to the system. Balancing priorities between scheduled and unscheduled utilization of the same communication network shared by many users is extremely difficult. The unscheduled utilization by paying subscribers normally receives top priority.

A hybrid technology which has been successfully applied is using the public telephone network to communicate with a data concentrator which subsequently uses a low-power RF link to communicate with different metering devices. A data

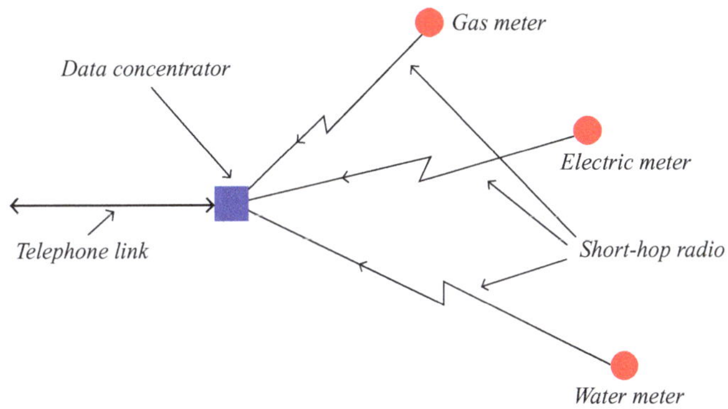

Figure 7.4. Diagram showing the telephone link to the data concentrator box and the radio frequency links to the various types of meters at customer's premises.

concentrator unit as shown in figure 7.4 at the customer premises may or may not be directly linked to the electric meter and is also linked to the gas and water meters by means of short hop radios. This unit collects information from water, gas and electric meters and the collected information can be retrieved and subsequently transmitted to the customer service computer via the telephone network on a scheduled basis. The primary node in this case is the data concentrator.

For large scale implementation, the communication system depends very much on the reliability and accessibility of the telephone network system. In addition, communication time and the use of available channels have to be shared with public users. Finding the correct location for the data concentrator and determining which metering devices can be linked to the data concentrator are not simple tasks. Access to many of the metering devices can only be determined by actual field measurements.

Obstructions to the RF link between the meter reading device and the data concentrator inadvertently installed by a customer may cause communication problems. Several experimental systems using fiber optic links to the data concentrators and remote switching devices have been tried. The end points for fiber optic links require power to be functional. An outage can cause the data concentrator to become de-energized and not functional. Unless an independent energy source is available to power the data concentrator and the end point of the fiber optic link, power outages will adversely affect the communication network.

7.4 Radio frequency technology

RF technologies use air space as their signal propagation medium and operate in the regulated or unregulated bands. The unregulated band was initially in the range of 1.0 GHz and above. Recently, the surge of broadband technology and its applications in the area of entertainment technology has triggered the Federal Communications Commission (FCC) to step in to regulate the frequency allocations.

The same air space is shared by many RF communications. Separation amongst users in the regulated band is controlled by the FCC and is accomplished by frequency allocation, transmitter power and modulation methods. A fixed network

Table 7.2. List of radio frequency bands currently government regulated and unregulated.

Frequency range (mHz)	Band designation	Regulatory treatment
150–70	150	Licensed
217–22	220	Licensed/auction
902–28	900	License free
1390–1435	1.4	Licensed/auction
2400–2483.5	2.4	License free
5725–5850	5.8	License free

consists of nodes with antennas that serve as repeaters and may or may not contain intelligent memory devices. In some radio technology these intelligent nodes also serve as data concentrators.

7.4.1 General characteristics of popular wireless operating frequencies

Many wireless advanced metering infrastructure (AMI) and distribution automation (DA) systems use radio spectra between 150 MHZ and 1500 MHZ. A more detailed breakdown of the common frequency bands and the regulatory treatment is shown in table 7.2.

7.4.2 One-way radio frequency technology

One-way RF technology has been extensively used in the USA for load management and capacitor bank control. High power transmitters and repeaters can reach a large number of remote transponders over large geographical areas using global commands. The outbound communication performance is not easily measured or monitored. Separation of non-respondents to outbound commands from the transponders that are not functioning is almost impossible from a large scale standpoint. To convert this system into a two-way system in order to expand its capability to handle new applications will require an increase in the number of lower power subnodes. Each of the large number of transponders at remote sites will require a power limited inbound transmitter to be retrofitted. All the nodes and subnodes or repeaters will have to be modified to handle two-way communication.

7.4.3 Short hop RF communication and the high power two-way RF trunking system

One system for meter reading uses short-hop RF between a data concentrator and several one-way RF transmitters retrofitted into the customer meters. Metering data are continuously transmitted by each transponder over a low power one-way RF link to the data concentrators. The data are received, collected and time-stamped by the data concentrator, which subsequently transmits the collected data through a high power RF trunking communication network to the service computer center as shown in figure 7.6.

The local transmitters at the transponders continuously sent metering data to the data concentrator and do not operate in a scheduled mode, hence collisions will occur.

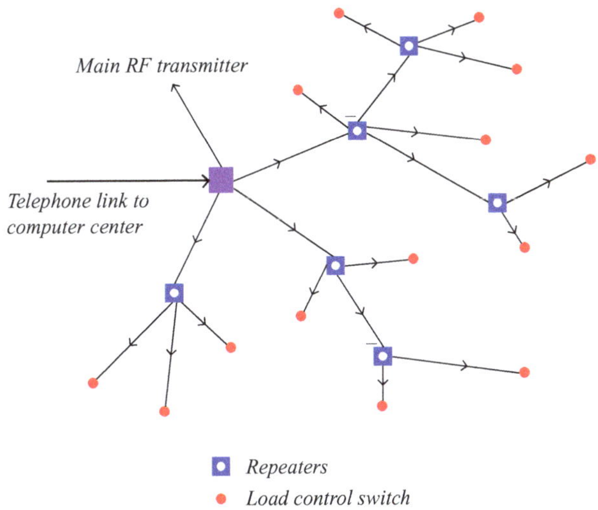

Figure 7.5. Diagram of a one-way RF communication network incorporating RF repeaters from the main transmitter to the various customers.

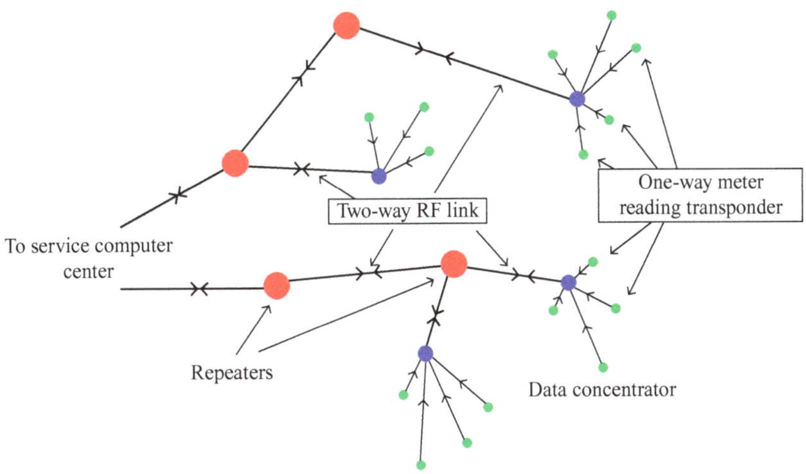

Figure 7.6. Diagram of a two-way RF network, linking data concentrators to one-way RF links to the meter reading transponders at the customer's premises.

Multiple repeats with random transmission delays will eventually cause all the data from the meters to be downloaded to the data concentrator. Time stamping of data occurs at the data concentrator.

When an outage occurs at the transponder, there is enough energy stored at the transmitter for a few communication retries. The latest meter reading prior to the loss of power may or may not be lost due to a collision. After a few transmission retries it is hoped that the data will be completely received by the data concentrator. Hence, local outages at single residences are detectable by the data concentrator.

A large area power outage resulting in loss of power for some intermediate nodes may cause operational problems. This is especially true when many transponders remain energized but the communication path to the service center computer is not available due to loss of power at some nodes and no other nodes can be used for alternative routing.

7.4.4 The low power multi-node two-way radio frequency

The low power two-way RF communication operates at very high frequencies. These frequencies are in the unregulated band and the communication system uses spread spectrum technology. At the operating frequency, the transmitter and receiver have to be within line of sight and cannot be too far apart (figure 7.7). Each node, upon receipt of a message from another node, has to send an acknowledgment to the sender. Otherwise, the sender will try to send the message again. This fact can cause some problems if a node is not functioning properly or severe communication problems exist between two nodes. The following example illustrates the type of problem that might occur.

Assume that transponder T2 sends a message to node ND7 and following a preprogrammed routing algorithm the message will subsequently be transmitted by node ND7 to ND5. But node ND5 has a problem and cannot receive the information from ND7. Hence ND7 never receives an acknowledgment from ND5 and node ND7 keeps trying to send the information to node ND5. Basically, ND7 is busy retrying and hogs that link and does not allow other transponders to use node ND7. To avoid this operational problem, by design ND7 is allowed only a few communication retries. If it fails to receive the acknowledgment from ND5, ND7 clears its memory and accepts new information from other transponders or other nodes for transmission to other nodes. At the central computer center, one has

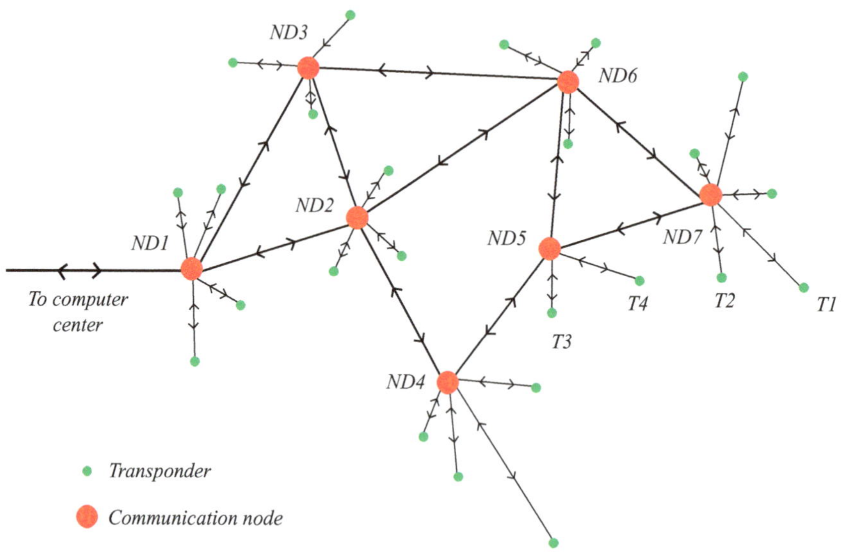

Figure 7.7. Diagram of a two-way multi-node RF network for meter reading.

absolutely no idea that there is a problem at node ND5 and that data from transponder T2 are lost. To overcome this problem, no single dedicated communication route should be used.

An automatic routing algorithm should try ND6 as an alternative destination. Sufficient retry capability should be available just in case ND6 is busy handling other nodes. From an infrastructure standpoint, the large numbers of nodes required makes it non-economical for use in rural areas, where customers are few and scattered over large geographical areas. In areas with high-rise buildings extra nodes are required to go around the buildings. Since these nodes require power to operate, standby power may be needed in case there is a power outage.

7.4.5 Broadband systems using the power line and other types of communication systems

Lately there has been a lot of publicity around successes using the medium voltage power line for broadband applications operating at very high baud rates. The costs of equipment and components for filtering, high voltage isolation, line conditioning, etc, per medium voltage feeder over the number of points served will be very high unless other non-utility applications can be found. From a cost-per-point basis, most of the cost is passed on to the customer. These customers sign up for services for personal use. These are primarily applications for internet services, real-time text, video, audio transmission, etc.

Using these systems for electric utility applications only may be cost prohibitive. Meter reading, load management applications, etc, do not require the very high baud rate that the broadband services can offer. None of the broadband PLC research and pilot programs using the utility power network are driven by automatic meter reading applications, distribution automation, etc. If the research and pilot system studies indicate that broadband power line (BPL) carriers are economically viable and can reach all customers served by the utility network, then typical utility functions may be able to time share using this technology.

The frequencies used are in the MHz range. Baud rates of 20 Mbit are being considered for power line carriers at the medium voltage level. In the low voltage network, in order to accommodate HDTV, baud rates of up to 200 Mbit s^{-1} are being considered.

If these frequencies are injected into the power line, every power outlet at residences, offices and industrial facilities will see these broadband carrier frequencies. It is not clear how they might affect computers operating at high frequency clock rates, local communication links such as portable telephones, short range RF communication devices, etc. If interference due to BPL carriers becomes noticeable to other devices, the government will eventually step in to regulate the frequency allocations.

Most BPL systems have similar architectures. Due to the low signal power used by BPL systems, repeaters may be required at certain intervals and bridging distribution transformers to reach the customers' premises requires special techniques. Some vendors use hardwired methods to bridge from the primary high voltage side node to the node located at the low voltage side of the distribution transformer.

All nodes are capable of generating and regenerating BPL signals. Another method used is to use WiFi equipment to send the signals to the customer premises which eliminates the need to have a node at each distribution transformer.

BPL falls under the jurisdiction of the FCC part 15, which covers all unlicensed applications. All users of the unlicensed spectrum should not cause harmful interference to others. It is not quite clear yet how these frequencies affect existing digital electronic devices. More field studies have to be performed to obtain a metric for possible damaging interference. Everything regarding broadband communications is still in a state of flux due to regulatory requirements at the municipal, state and federal government levels. A clearing-house to provide information on all existing BPL projects is made available by the FCC. In a public notice on October 15 2006, the FCC certified the United Telecom Council (UTC) as the Access Broadband over Power Line (BPL) Database manager. All BPL operators have to post information about their project to the database. Public access to search the information is at www.bpldatabase.org.

The fiber optic communication network has also been considered for typical electric utility type applications. For meter reading, a large number of taps and junctions are required. Each of these taps or nodes requires power. If the electrical supply to a node is out due to a localized electric utility network outage, then the tap becomes dysfunctional. To install a backup power supply at each node is not economical. For high-speed point-to-point trunking systems, a fiber optic system is very reliable and can be very economical.

Another system currently under consideration is the use of two-way satellite communication between the communication server and the data concentrators. The use of the public digital cellular telephone systems for meter reading applications has been discussed. The economics of this method are not clear yet and no sufficiently large operating systems are currently available which can be used for reference and discussion.

The following section lists the new RF network systems that are being considered for the transmission of intelligence and entertainment. Some suggestions are being made to use them as links between the medium voltage network and the low voltage network.

7.4.6 Recommended practices for using wireless data communications in power system operations

These recommended practices describe and make recommendations on the functional, performance, security and on-site testing issues related to using wireless data communication technologies in different aspects of power system operations, including within electric power substations, in underground vaults, along transmission and distribution circuits, within generation and distributed generation plants, for customer electrical and metering equipment, and other electric power environments.

Although recognizing that existing wireless media, such as microwave, MAS radio, spread-spectrum radio, VSAT satellite systems, and proprietary radio-based systems, could also be used for some applications, these recommended practices focus on the newer wireless data communication technologies.

WiFi, the most popular wireless standard for networking computer systems, has the following basic characteristics.

- A multi-user configuration.
- The IEEE 802.11b data rate is 11 Mbps.
- The IEEE 802.11 g data rate is 54 Mbps.
- The frequency band is the 2.4 GHz band.
- It has a range of 100–50 feet.
- The equipment and WiFi systems/access points are usually privately owned.

Bluetooth is used in cell-phones, personal digital assistants (PDAs), and other mobile wireless devices, primarily for communicating with computers, intelligent electronic devices (IEDs), headsets, hands-free systems, and other gadgets. Bluetooth has the following characteristics.

- Point-to-point links.
- A very short range of only 33 feet (approximately 10 m).
- The frequency band is the 2.4 GHz band.
- A relatively low data rate of 1.5 Mbps.
- The equipment and Bluetooth systems are privately owned.

ZigBee, based on IEEE 802.15.4, defines low-rate, very low duty cycle, wireless personal area networks often termed 'meshed networks' as opposed to point-to-point. ZigBee builds upon the 802.15.4 standard to define application profiles that can be shared among different manufacturers to provide system-to-system interoperability (open standard). This effort is still a work in progress, although it is of great interest to industries (such as the power industry) that have extensive sensor networks. An additional advantage is that a ZigBee end device can be configured for very low power consumption, staying in a sleep mode until activated. ZigBee has the following characteristics.

- A multi-user configuration.
- An ultra-low output power (in the mW range).
- Frequency bands:
  - 863–70 MHz in Europe;
  - 902–28 MHz in the USA;
  - 2.4–2.4385 GHz worldwide.
- The range between devices is 30–300 feet.
- A low data rate of <250 kbps.
- There are three types of ZigBee devices:
  - ZigBee end device (ZED), with the ability to communicate with other nodes on the network and to interface the end device to the network;
  - ZigBee router (ZR), which acts as an end device and also as an intermediate router to other nodes; and

  - ZigBee coordinator (ZC), which acts as the control of the network and provides a bridge or gateway to other networks (utilities' power line carrier, wireless AMI, etc).
- A high availability due to meshed network configuration.
- The equipment and ZigBee systems are usually privately owned.

WiMax (IEEE 802.16) wireless technologies have the primary focus of enabling a wireless alternative for cable, digital subscriber line (DSL) and T1 communication channels for consumer last-mile access to the Internet, including high-speed data, voice over IP (VoIP), video on demand (VoD) and backhaul for IEEE 802.11 local area networks (LANs). WiMax addresses the 'first-mile/last-mile' connection for longer distances and faster rates. The characteristics of WiMax are as follows.

- A multi-user configuration.
- A range of 5–30 miles (5 miles being more likely).
- Data rates of 45–75 Mbps.

The *cell-phone data standard*, general packet radio service (GPRS), is part of the GSM effort to create a common European mobile telephone standard for a pan-European mobile cellular radio system (and now worldwide). The resulting mobile telephone standard allows cell-phone users to 'roam' across many cell-phone systems and between most countries worldwide. New generations of cell-phone technologies, called 2.5 G, 3 G and 4 G, have been deployed in certain countries or are still under development. Cell-phone systems are owned by telecommunication providers. Their characteristic are as follows.

- A multi-user configuration.
- GPRS is commonly used for data, with rates of 30–80 kbps typical.
- EDGE (enhancement to GPRS) provides rates of 160–236 kbps.
- The range is wherever cell-phone coverage is available!

IOP Publishing

New Technologies for Smart Grid Operation

Sioe T Mak

Chapter 8

Comparisons of technologies

The degree of complexity for implementing future added-value functions and applications to existing systems varies amongst the different technologies. The following lists are an attempt to provide a degree of measure in terms of complexity amongst the various technologies and whether some of the added-value functions as listed in chapter 5 can be implemented.

In this attempt to compare technologies, the applications of using cellular mobile systems, power line carrier systems for home automation and broadband systems are purposely left out due to a of lack of good information. Pilot projects are numerous and large-scale implementations have slowly materialized. Power line carrier systems for low voltage networks, which are very popular in Europe, have not sufficiently penetrated the USA, the reason for which has been explained in chapter 7.

Table 8.1. One-way systems using powerline communication.

Functions	One-way outbound	One-way inbound
1. Advanced metering and time-of-use.	Requires inbound capability. Requires metering transponders.	Requires outbound capability. Replacement of metering transponders required.
2. Gas and water metering.	Requires inbound capability. Subnetwork required.	Subnetwork required.
3. Service disconnect.	Possible, for status indication inbound capability is required.	Not possible unless outbound capability is installed.
4. Load management.	Initial intended application, but no status verification.	Not possible.
5. Alarm.	Possible, general system alarm.	Triggered by local event, time delay.
6. Outage management.	Not possible without inbound capability.	Possible for outage mapping.
7. Integrated volt-VAR control.	Possible with help of SCADA.	Not possible.
8. Remote switching.	Possible with help of SCADA.	Not possible.
9. Feeder load balancing.	Possible with help of SCADA.	Not possible.
10. Power quality monitoring.	Not possible.	Possible.

doi:10.1088/978-0-7503-1158-8ch8

Table 8.2. Two-way power line communication.

Functions	Medium frequency	Low frequency
1. Advanced metering and time-of-use.	Improve immunity to network changes and communication spill-over.	Increase in data throughput required.
2. Gas and water metering.	Requires subnetwork communication.	Requires subnetwork communication.
3. Service disconnect.	Possible.	Possible.
4. Load management.	Possible.	Possible.
5. Alarm.	Possible.	Possible.
6. Outage management.	Possible, dead line communication should not be required.[a]	Possible, dead line communication should not be required.[a]
7. Integrated volt-VAR control.	Implementation is simple.	Implementation is simple.
8. Remote switching.	Implementation is simple.	Implementation is simple.
9. Feeder load balancing.	Implementation is simple.	Implementation is simple.
10. Power quality.	Power quality problems may impair communication reliability.	Power quality problems may impair communication reliability.

[a] A 'dead line' is part of the distribution network which is de-energized and completely isolated from the rest of the network.

Table 8.3. Radio frequency communication.

Functions	One-way outbound	Two-way communication
1. Advanced metering and time-of-use.	Requires inbound capability.	Possible.
2. Gas and water metering.	Requires inbound capability.	Possible.
3. Service disconnect.	Possible, for status indication inbound capability is required.	Possible.
4. Load management.	Initial intended application but no status verification.	Possible.
5. Alarm.	Possible, general system alarm.	Triggered by local event possible.
6. Outage management.	Not possible without inbound capability.	Possible, strongg relational link between communication network and distribution network required.
7. Integrated volt-VAR control.	Possible with help of SCADA.	Implementation is simple.
8. Remote switching.	Possible, status verification requires inbound capability.	Implementation is simple.
9. Feeder load balancing.	Possible with help of SCADA.	Implementation is simple.
10. Power quality monitoring.	Possible, requires inbound capability.	Possible.

Table 8.4. Hybrid telephone/RF trunking + local short hop radio technology.

Functions	One-way local short hop radio transponders to data concentrator	Two-way local short hop
1. Advanced metering and time-of-use.	Feasible, large intelligent data concentrator link required.	Hourly data can be locally stored at the transponder.
2. Gas and water metering.	Generic implementation.	Redundant RF two-way network unless future time-of-use is considered.
3. Service disconnect.	Not possible.	Possible.
4. Load management.	Not possible.	Possible.
5. Alarm.	Not possible.	Possible.
6. Outage management.	Possible, relationship of transponder location to distribution network required.	Possible, relationship of transponder location to distribution network required.
7. Integrated volt-VAR control.	Not possible.	Possible.
8. Remote switching.	Not possible.	Possible.
9. Feeder load balancing.	Not possible.	Possible.
10. Power quality monitoring.	Possible.	Possible.

Table 8.5. Other technologies.

Function	Fiber optic[a]	Satelite radio[a]
1. Advanced metering and time-of-use.	Feasible, total saturation not expected soon.	Inbound to base-station required.
2. Gas and water metering.	Subnetwork required.	Inbound to base-station required.
3. Service disconnect.	Possible.	Inbound to base-station required.
4. Load management.	Requires large saturation of the network.	Possible.
5. Alarm.	Possible.	Possible.
6. Outage management.	Selectively possible.	Inbound to base-station required.
7. Integrated volt-VAR control.	Possible.	Possible.
8. Remote switching.	Selectively possible.	Possible, status verification requires inbound.
9. Feeder load balancing.	Possible.	Possible.
10. Power quality monitoring.	Possible.	Inbound to base-station required.

[a] Only economical if coupled to broadband applications for which customers are willing to pay for subscriptions.

IOP Publishing

New Technologies for Smart Grid Operation

Sioe T Mak

Chapter 9

New issues facing smart grids

Automatic meter reading is slowly becoming reality due to advances in communication technology, database management systems and the introduction of smart meters. Smart grid applications are being explored for optimizing and improving the reliability of the distribution network to support generation, energy delivery and customer satisfaction. Demand response, outage management, asset management, distribution automation, novel rate designs, etc, are applications that are already being considered as smart grid functions. It is still too early to assess which way the industry is moving in terms of priority of efforts, primarily because of uncertainties about the capabilities of existing technologies to enable the implementation of smart grid functions. During this unsettled period, new concerns are looming over the horizon. Electric vehicles (EVs) and hybrid EVs are gaining popularity and coming down in price and soon will be within reach of the general population. Distributed generation, such as photovoltaic (PV) cells, wind power generation or a combination of both, is gaining popularity with small businesses and in the suburban areas of the USA. All of these add another aspect of complexity to the operation of the distribution network. The impact on demand response, revenue metering, power quality, and the monitoring and control of faults in the network will be discussed in the following sections.

9.1 Electric vehicles and hybrid electric vehicles

One liter of gasoline contains about 9.7 kWh of energy and internal combustion has an efficiency of only 25%. An EV using a battery and electric motor combination has an efficiency of about 75%. A simple example showing the cost differences between the two types of engines will clarify this comparison.

1. A gallon of gasoline contains approximately 33.0 kWh of energy. Assume a 25% efficient internal combustion engine for an automobile which travels 15 miles for one gallon of gasoline. The effective kWh used to move the car is $0.25 \times 33.0\,\text{kWh} = 8.25\,\text{kWh}$.

doi:10.1088/978-0-7503-1158-8ch9

2. A 75% efficient battery and electric motor combination requires, for the same number of miles, a total energy intake of (100/75) × 8.25 kWh, which is 11.0 kWh.
3. Assume that the cost of charging the battery is \$0.10 kWh^{-1}, the energy cost is \$1.10 per 15 miles of driving.
4. Assume a price of \$3.30 per gallon of gasoline, the energy cost to drive 15 miles is \$3.30.

Hence for a battery and electric motor combination, electric energy is less expensive provided its cost is approximately less than three times of that of gasoline.

EVs draw their charging currents from the service voltage network. There is some talk in the industry about multi-tier rate structures for the level of charging used (fast and slow charging). Charging at 1.4 kW at 120 V takes approximately 8–10 h and at 3.3 kW at 240 V the charging time is approximately 4–5 h. During the charging cycle, the current drawn from the service voltage circuit is pulsating and as the battery is charged the pulse at each half cycle of the power voltage becomes smaller in magnitude with a progressively narrower base per pulse. The total harmonic distortion (THD) of the current changes as the battery nears full charge.

If a few EV charging systems are connected to a distribution transformer, the distribution transformer voltage will also be distorted. Since the rectified current uses the local house voltage for charging, with many homes with EV charging systems served by a distribution transformer, the local voltages at the homes may be phase shifted from each other. Hence the pulses from the homes are not necessarily in phase. The net THD of the current is not identical to the THD of the current at each home and the net effect on the supply voltage from the distribution transformer is more complex. What effects will this have on the losses at the network and the distribution transformer? If the charging rectifier is unbalanced (one rectifier diode dysfunctional), dc current will be drawn from the transformer. How will all this impact the accuracy of revenue metering and kVAh metering?

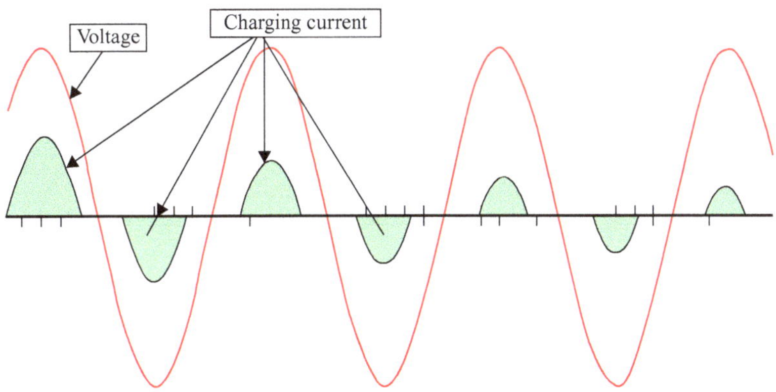

Figure 9.1. The possible shape of the battery charging current from the source using a bridge rectifier.

9.2 Photovoltaics and wind generators

9.2.1 Photovoltaics

A typical PV array and windmill combination is shown in figure 9.2. The windmill generator and the PV array both provide a dc output to the storage batteries. To overcome ac synchronization problems, the whole dc generation and battery storage assembly are coupled to the electric utility's ac source and to the internal load inside the home through dc to ac inverters. A disconnect at the utility control

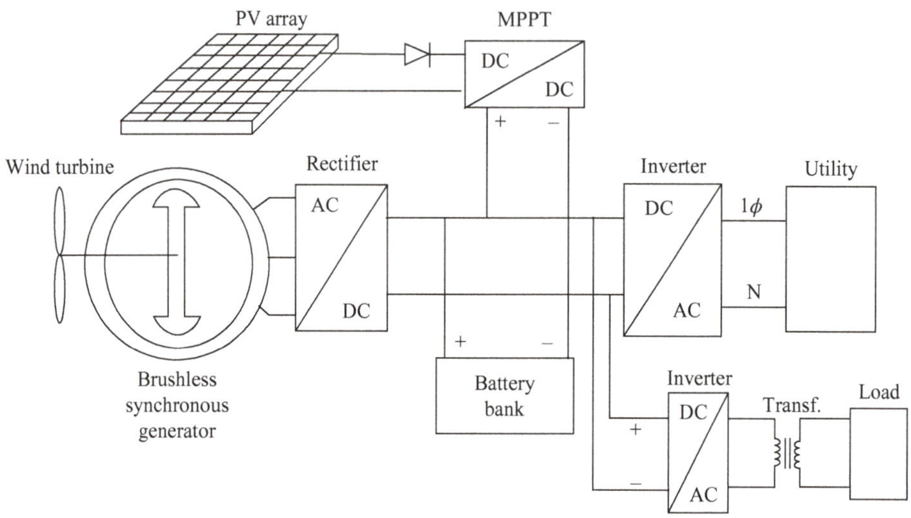

Figure 9.2. A schematic diagram of a typical connection from a PV array and a windmill driven generator to a storage battery and the links to the electric utility distribution network and customer loads.

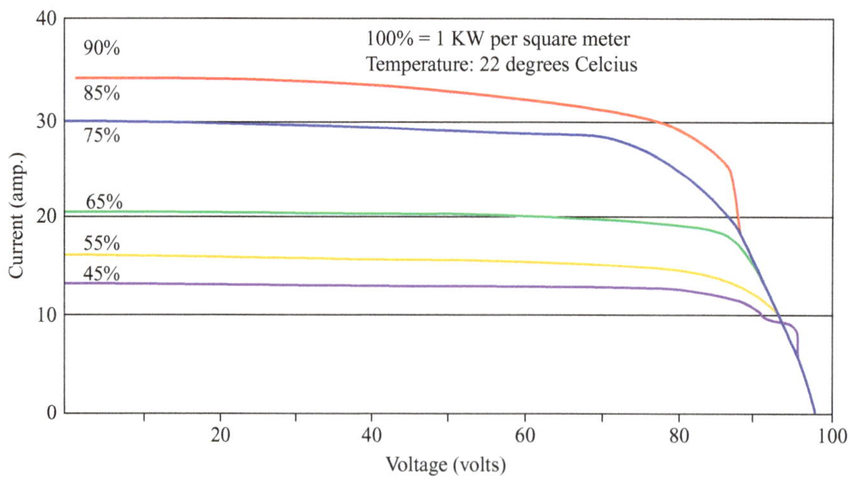

Figure 9.3. Voltage/current characteristics of the solar generator at various irradiance levels.

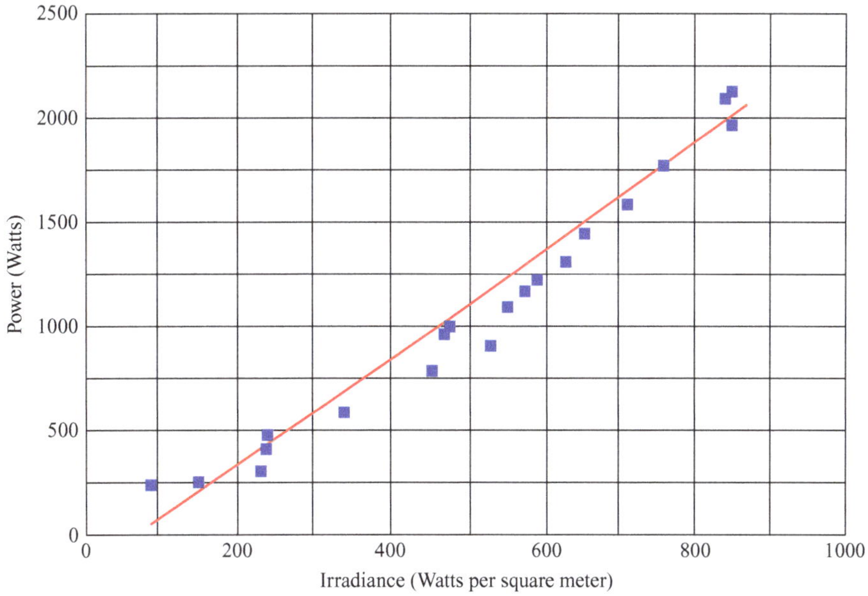

Figure 9.4. PV output power versus irradiance taken at 20 °C.

panel will cause all loads to be energized by the battery system. A typical PV array consists of between 20 and 25 modules. Each module is built up of 36 individual 600 mV solar cells. A typical module rating as recommended by the manufacturer is 53 W.

9.2.2 Wind turbines

A typical 1.5 kW upwind three-blade horizontal axis turbine coupled to a permanent magnet brushless synchronous generator characteristic curve is shown in figure 9.5. The turbine aligns itself to the wind direction until it reaches the furling wind speed of $13\,m\,s^{-1}$. Upon reaching the furling speed, the turbine will face out of the direction of the wind.

9.2.3 Fault current issues

Small distributed generators at single homes or smart homes are metered separately. However, when they generate excess power, more than the customer needs, this power is delivered to the utility network and monitored by the electric utility meter as reversed power usage. If a fault occurs on the distribution feeder supplying the power to customers with or without distributed generators, the fault current will be supplied by all sources on the feeder. These contributors to the fault current are the utility distribution substation and all the distributed generators connected to the network served by the feeder. Hence one can expect fault currents to flow through the utility meter at the customer premises. What effect does this have on the meter?

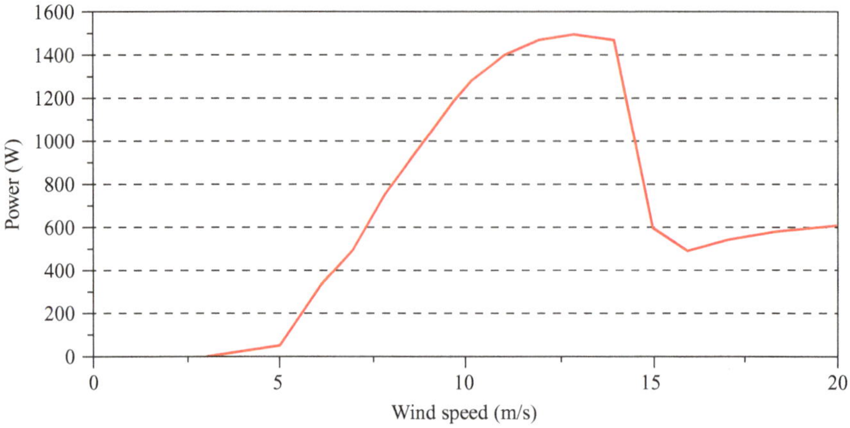

Figure 9.5. Power output versus wind velocity of a typical 1.5 kW upwind, three-blade horizontal axis turbine coupled to a permanent brushless synchronous generator.

9.2.4 Islanding issues

After the fault on the feeder is cleared (disconnected from the distribution substation bus), the fault current contributions by the individual local generators may or may not disconnect the generators from the network. This may become a problem called *islanding*. With the possibility of induction motors with large inertia acting as induction generators and the distributed local generators interconnected by the isolated network, power swings between the induction generators and the distributed local generators will happen. There will be power flow going back and forth through the utility meters of customers who own local generation. This will affect the meter reading.

IOP Publishing

Sioe T Mak

Chapter 10

Conclusions

In this book, we have seen that terms such as 'high baud rate', 'retries', 'monitor outages' and '100% communication performance' can be misleading unless they can be tied to infrastructure design, system operation, data throughput per unit time, database design, etc. The essential building blocks which constitute the system architecture have been briefly described and discussed.

An attempt has been made in this book to show the synergism which must exist amongst various applications and functions in order to make a system operate in an optimal fashion. It is also necessary to implement the outage management function in both the power delivery network and the communication network and provide the necessary links between the two network databases. There are physical links amongst the paths and nodes between both networks. In addition to supporting the maintenance and repair department in the speedy location and repair of faults and the restoration of segments of networks, the availability of information on the topological and geographical locations of protective devices and major nodes will greatly enhance the system operation.

Limitations of space do not permit a more detailed discussion of the requirements for the implementation of advanced metering and other distribution functions. I hope that the fundamental issues described and discussed in this book provide the reader with some guidance as to what questions to ask about the communication systems for utility applications which are now available in the market.

Further reading

Arseneau R and Filipsky P S 1988 Application of a three phase nonsinusoidal calibration system for testing energy and demand meters under simulated field conditions *IEEE Trans. Power Deliv.* **3** 874–9

Arseneau R *et al* 1996 Practical definitions for powers in systems with nonsinusoidal waveforms and unbalanced loads: a discussion *IEEE Trans. Power Deliv.* **11** 79–101

Arseneau R, Heydt G T and Kempker M J 1997 Application of IEEE standard 519-1992 harmonic limits for revenue billing meters *IEEE Trans. Power Deliv.* **12** 346–53

doi:10.1088/978-0-7503-1158-8ch10

Baghzouz Y and Tan O T 1985 Harmonic analysis of induction watthour meter performance *IEEE Trans. Power Appar. Syst.* **PAS-104** 399–406

Devaney T 1995 Successful communication techniques for distribution systems *DA/DSM ASIA (Singapore, 1995)* conference paper

Ehrenreich D and Liberman S 1995 Cost benefits resulting from use of integrated communication for distribution automation *DA/DSM ASIA (Singapore, 1995)* conference paper

Emanuel A E 1990 Powers in nonsinusoidal situations, a review of definitions and physical meaning *IEEE Trans. Power Deliv.* **5** 1377–89

Ferrero A, Menchetti A and Sasdelli R An approach to energy metering in the presence of distortion, asymmetry and unbalance

Goldberg S and Horton W F 1987 Induction watthour meter accuracy with nonsinusoidal currents *IEEE Trans. Power Deliv.* **2** 683–90

IEEE 1986 *Load Management Book* (Piscataway, NJ: IEEE)

IEEE 1993 *Recommended Practices and Requirements for Harmonic Control in Electrical Power Systems, IEEE Std 519-1992* (Piscataway, NJ: IEEE)

Lin W M, Tsay M T and Wu S W 1996 Application of geographic information system to distribution information support *IEEE Trans. Power Syst.* **11** 190–7

Mak S and Radford D 1995 Design considerations for implementation of large scale automatic meter reading systems *IEEE Trans. Power Deliv.* **10** 97–103

Mak S and Radford D 1996 Communication system requirements for implementation of large scale demand side management and distribution automation *IEEE Trans. Power Deliv.* **11** 683–9

Mak S T 1987 The random failure statistics and replacement schedules of remote devices for load management *IEEE Trans. Power Syst.* **2** 451–7

Mak S T 1999 Synergism between intelligent devices and communication systems for outage mapping in distribution networks *CIRED 15th Int. Conf. Electricity Distribution (Nice, France, 1–4 June 1999)* (http://www.cired.net/publications/cired1999/papers/3/3_p2.pdf)

Mak S T 2001 Power delivery infrastructure differences and their impacts on different types of power line communications for automatic meter reading *CIRED 16th Int. Conf. Exhibition on Electricity Distribution (Amsterdam, 18–21 June 2001)* vol 4 (Piscataway, NJ: IEEE)

Mak S T and Hilleary T N 1995 A power frequency communication system designed for utility application *DA/DSM ASIA (Singapore, 1995)* conference paper

Mak S T and Radford D 1995 Added utilization costs associated with different communication architectures for distribution automation and demand side management *IEEE Trans. Power Deliv.* **10** 104–8

Makram E B, Wright C L and Girgis A A 1992 A harmonic analysis of the induction watthour meter's registration error *IEEE Trans. Power Deliv.* **7** 1080–8

McDonald J R, Burt G M, McArthur S D J, Bell S C and Elders I M 1995 Proposed architecture for integrated decision support functions within second generation distribution management systems *DA/DSM ASIA (Singapore, 1995)* conference paper

McEachern A, Grady W M, Moncrief M A, Heydt G T and McGranagha M 1995 Revenue and harmonics: an evaluation of some proposed rate structures *IEEE Trans. Power Deliv.* **10** 474–81

McEachern A, Heydt G T, Harbaugh W, Nash H, Hiatt E, Fauquemberque P, Mahlatsi T and Smith R 1996 Power quality and rate structures—should they be combined? *IEEE PES Winter Meeting (Baltimore, MD)* panel session

Modzelewsky T J and Osmansky A M 1993 Relating precise measurements of a solid state meter to economics through field study *EEI/AEIC Presentation (Indianapolis, IN, 20–23 September 1993)*

Munday M 1993 Current sensing technology *EEI/AEIC Presentation (Indianapolis, IN, 20–22 September 1993)*

Newbury J 1996 Communications field trials for total utility metering *IEEE Trans. Power Deliv.* **11** 690–5

Rabi V A *et al* 1986 *Proc. Int. Load Management Conf. (Chicago, IL, 9–11 December 1985)* (Palo Alto, CA: EPRI) EPRI-EM-4643, Project 1940-15

Stoll G 2007 Characteristics of wireless AMI and DA operating frequencies *UTC J.* (special issue)

Lightning Source UK Ltd.
Milton Keynes UK
UKHW05n0614180618
324301UK00003B/83/P

9 780750 311595